Eden Revisited

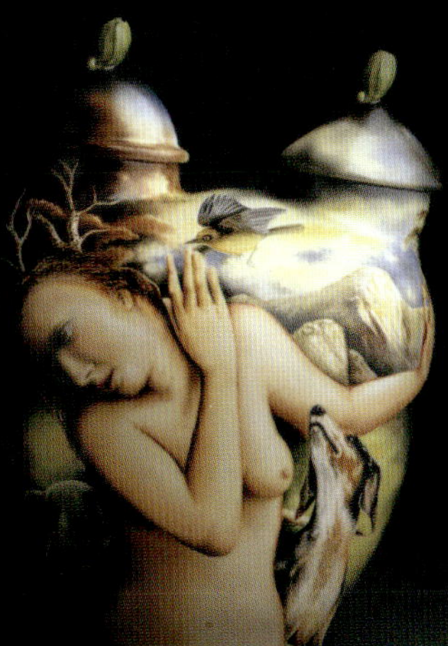

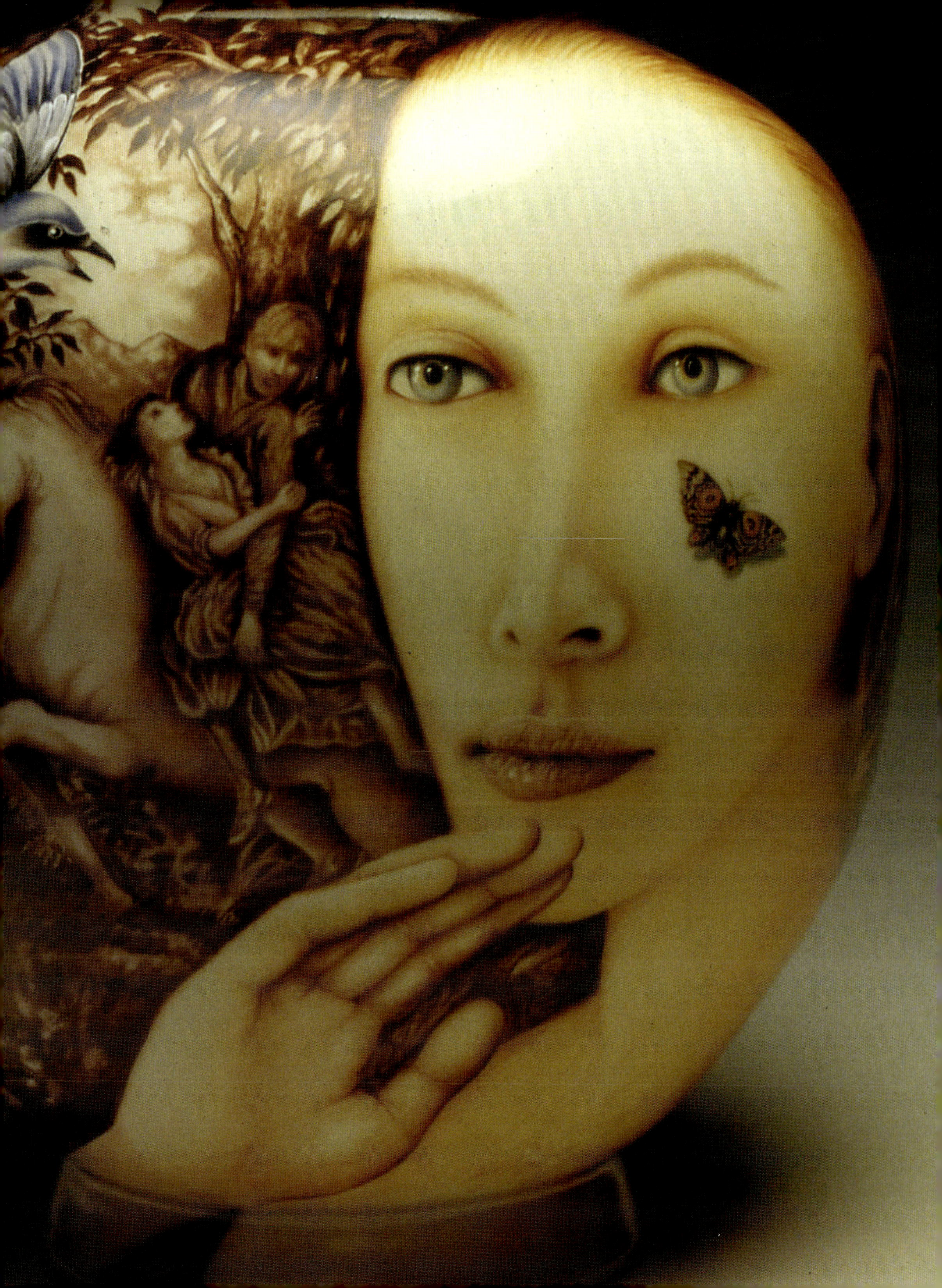

Eden Revisited

The Ceramic Art of Kurt Weiser

Edited by

Peter Held

Essays by

Ulysses Grant Dietz

Peter Held

Edward Lebow

Herberger College of the Arts
ARIZONA STATE UNIVERSITY

This book was published in conjunction with an exhibition organized and presented by the Arizona State University Art Museum Ceramics Research Center.

Arizona State University Art Museum
Herberger College of the Arts
Tenth Street and Mill Avenue
Tempe, Arizona 85287-2911
tel. 480.965.2787 fax 480.965.5254
http://asuartmuseum.asu.edu

Eden Revisited: The Ceramic Art of Kurt Weiser
Exhibition Touring Schedule

November 1, 2007–January 6, 2008
Museum of Contemporary Craft
Portland, Oregon

February 12–April 19, 2008
Bellevue Arts Museum
Bellevue, Washington

May 29–August 24, 2008
Holter Museum of Art
Helena, Montana

September 30–December 21, 2008
Nora Eccles Harrison Museum of Art
Logan, Utah

February 14–May 16, 2009
Arizona State University Art Museum
Tempe, Arizona

June 26–September 13, 2009
Erie Art Museum
Erie, Pennsylvania

January 22–March 20, 2010
Society for Contemporary Craft
Pittsburgh, Pennsylvania

April 29– June 27, 2010
Southwest School of Art and Crafts
San Antonio, Texas

Catalogue produced by
Arizona State University Art Museum

Arizona State University Art Museum

Produced for Arizona State University
by Perpetua Press, Santa Barbara, CA

Edited by Letitia Burns O'Connor
and Kathy Talley-Jones

Designed by Dana Levy

Printed in China by Asia Pacific Offset

ISBN 0-9777624-5-9
Distributed by
Arizona State University Art Museum
P.O. Box 872911
Tempe, Arizona 85287
(480) 965-2787
www.asuartmuseum.asu.edu

Photo credits:
All photographs by Craig Smith, excluding the following, courtesy of:

Kurt Weiser, pp. 11, 12, 13, 14, 15, 17, 29, 80, 81
Archie Bray Foundation for the Ceramic Arts,
pp. 20 (bottom), 24, 80 (right)
Anthony Cuñha 18, 19, 23, 31, 42, 43
The Newark Museum, 25, 26, 27
Garth Clark Gallery, Noel Allum 22, (left)
Richard Gehrke, 32
Ken Howie Studios, 40
McGraw-Hill, 37
National Museum in Krakow, Poland, 38
National Gallery of Art, Washington, DC, 38
Scott Swistowitz 40 (bottom)

COVER: Kurt Weiser. *Abduction Jar*, 1999. Plate 10 (detail)
PAGE 1: Kurt Weiser. *Fidelity*, 2000. Plate 11
PAGE 2: Kurt Weiser. *Abduction Jar*, 1999. Plate 10 (detail)
PAGE 5: Kurt Weiser. *Europa*, 2000. Plate 20 (detail)
PAGE 41: Kurt Weiser. *Red Leaf*, 1998. Plate 9 (detail)

Library of Congress
Cataloging-in-Publication Data

Eden Revisited: The Ceramic Art of Kurt Weiser/edited by Peter Held; essays by Ullyses S. Grant . . . [et al].
p. cm/
Catalog of an exhibition of the same name organized by Arizona State University Art Museum
includes bibliographical references and index
ISBN 0-9777624-5-9 (hardcover: alk. paper)
1. Kurt Weiser Exhibitions.
2. Ceramics—20th century Exhibitions.
I. Title: Kurt Weiser. II. Held, Peter III. Ullyses S. Grant. IV. Arizona State University Museum. V. Title

Contents

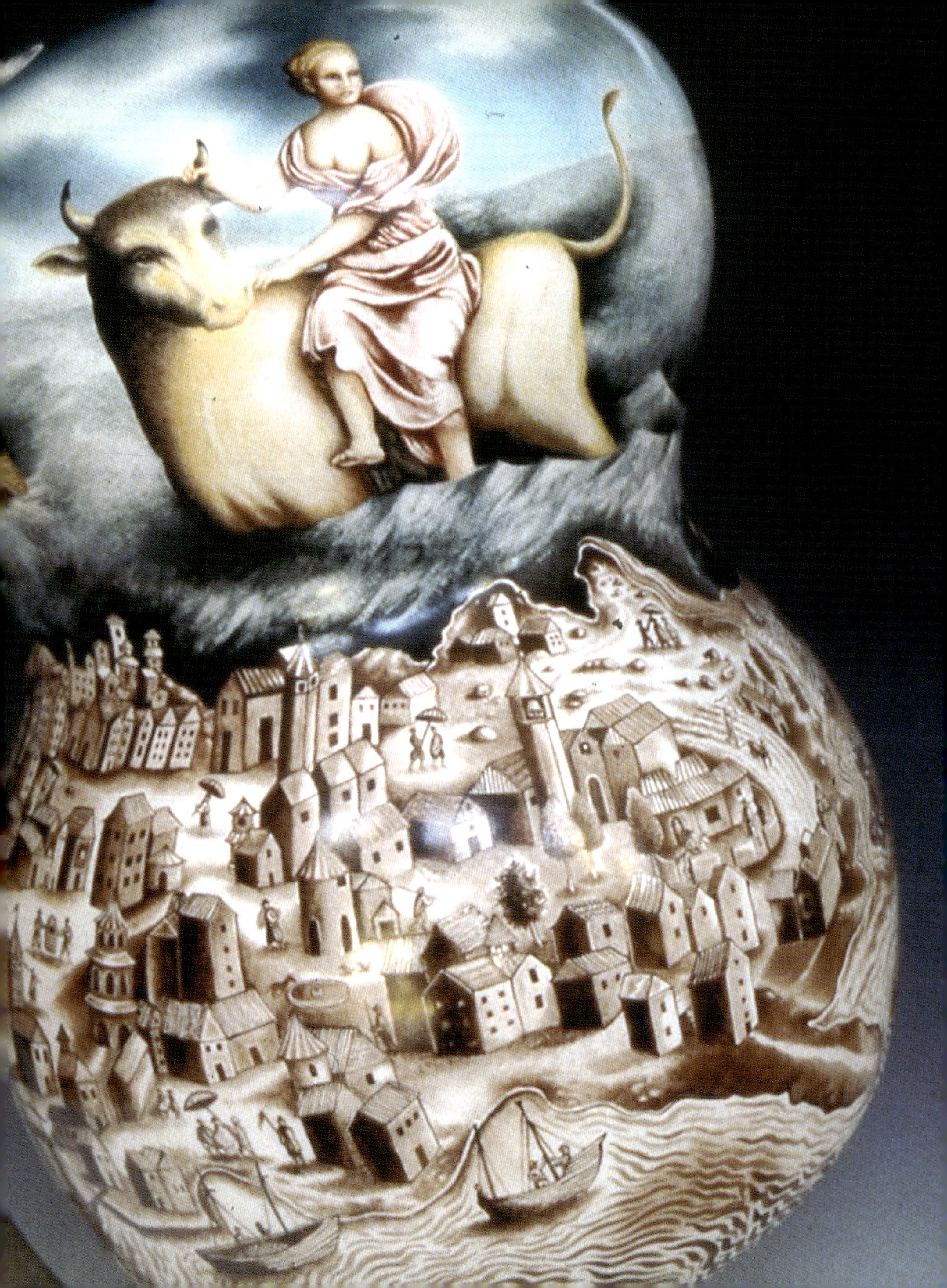

Foreword

Kwang-Wu Kim

Dean, Herberger College of the Arts

Arizona State University

Ceramics artist Kurt Weiser arrived at Arizona State University in 1989. Since that time, he has shared his vast knowledge of ceramics with students and embarked on new directions in his work. His appointment in 1999 as the first Regents' Professor in the School of Art honors him as a distinguished artist of national and international distinction and a teacher of exceptional ability.

Eden Revisited is the third in a series of monographic projects undertaken by the Arizona State University Art Museum Ceramics Research Center. The publication and the retrospective exhibition that it complements were made possible through the generosity of the Windgate Charitable Foundation, and the Herberger College of the Arts is most grateful for the recognition that this support demonstrates. *Eden Revisited* was realized not only by the hard work of Curator of Ceramics Peter Held, but also by his long association with Kurt Weiser. Finally, the publication and exhibition could not have happened without Kurt, who placed his trust in Peter and the center to present his work in a survey that reveals his evolution as an artist of international stature.

The ASU Art Museum has a long-standing commitment to craft-based media within a broader contemporary art context. Its mission is to serve a diverse community of artists and audiences through innovative programming that is both interdisciplinary and educational. Exhibitions, education programs, and publications are designed to engage viewers with art that is relevant to their lives.

This exhibition reflects the momentum of the museum's Ceramics Research Center, which is a national and international destination point for the hands-on study and enjoyment of ceramics. It combines an exhibition gallery with open storage and a research archive. The exhibition space provides a framework for focused exhibitions while the open storage allows the public access to a large portion of the permanent collection of more than four thousand objects. The center is a teaching and research facility, an educational component of the Herberger College of the Arts. The collection seeks to embrace representative works by artists,or subject matter that reflects the social, cultural, and historical activities of the world. The international holdings demonstrate the full range of technique, aesthetic approaches, and possibilities within the medium.

Under the leadership of University President Michael Crow, the "New American University" is being pioneered at ASU. His goal is to advance the development of educational models that reflect changing realities and place our students on a path to success in an ever-more complex world. The need for new models, thinking, and practices is profound in the arts.

While celebrating our historic traditions, we embrace a future in which new technologies, media, hybrid combinations of the traditional arts, and the growing influence of world cultures will all play their part in determining the content and nature of our work. At the Herberger College, we have committed ourselves to the challenge of leading with excellence while simultaneously creating wide access to the arts.

At the heart of all great institutions are individuals of extraordinary talent and imagination. We are pleased to be able to honor Kurt Weiser, a distinguished member of our faculty, whose exceptional gift and whose dedication to his students and his art help us realize our institution's future.

Introduction and Acknowledgments

Peter Held
Curator of Ceramics
Ceramics Research Center
Arizona State University Art Museum

It's always a pleasure for a curator to work on an artist's exhibition that demonstrates the talent and ingenuity of someone he or she holds in high regard. I've been an admirer of Kurt Weiser's ceramics since we first met at the Archie Bray Foundation for the Ceramic Arts in 1977, when Kurt became resident director. I was amazed at his technical prowess and disciplined work ethic from the start. Whenever his work took a major shift, he nimbly developed new clay bodies and glazes to realize his artistic vision. He is an unassuming genius, never satisfied to follow the pack, content to drift from the ceramic mainstream. Examining his creative output, we encounter an iconoclast pursuing dreams, memories, and desires in a centuries-old medium.

As an Arizona State University School of Art Regents Professor, Kurt Weiser has not only garnered recognition exhibiting his ceramic works at prestigious galleries and museums worldwide, but through teaching and example, enabling his students to follow similar distinguished paths. Redefining his artistic career over three decades of creative excellence as a skilled potter, master draftsman, and accomplished china painter, Kurt, an American original, has created a singular body of work that provides insight into the mind.

My deepest appreciation and gratitude is extended to Kurt and his wife Christy for their tireless efforts in making this exhibition and publication possible. They have been instrumental in my research efforts, providing necessary documentation at critical turns. My hope is that a wider audience will come to appreciate Kurt's ingenuity and creativity over a career that spans three decades pursuing artistic excellence.

I would also like to thank the two contributing essayists for their reflective observations of Weiser's work. Edward Lebow is one of the most thoughtful and provocative writers in the ceramics field. Living in Tempe, Arizona, he has witnessed the artist at close range and shares intimate knowledge of the artist garnered over a fifteen-year relationship. Ulysses Grant Dietz, Curator of Decorative Arts at the Newark Museum, has been recognized expert in the field of decorative arts for more than two decades. At ease with both historical and contemporary works in clay, his essay provides an insightful understanding of china painting.

I am also indebted to Dana Levy and Tish O'Connor of Perpetua Press, Santa Barbara, who have produced hundreds of important museum catalogues worldwide. Generous collaborators, they have been sensitive in showcasing Kurt Weiser's work to its best advantage.

Craig Smith from Phoenix, Arizona, has photographed the artist's work for many years with continued sensitivity and technical proficiency. We are also indebted to Anthony Cuñha, a Los Angeles-based photographer, who photographed the works in located in Montana. Through these incredible efforts, the artist's work has been interpreted to resonate in a different medium.

Eden Revisited: The Ceramic Art of Kurt Weiser will travel to eight additional museums on its national tour. I would like to acknowledge and graciously thank the museums and their staff for enthusiastically embracing the exhibition and attendant educational programs. The touring schedule, listed on page 4, will provide an opportunity for a broad audience to become familiar with the full range of Weiser's ceramics.

Fortuitously, the Windgate Charitable Foundation initiated its Artist's Exhibition Series granting program in 2005. The award for the Kurt Weiser project was comprehensive, providing the necessary funds to fully realize the exhibition and an eagerly awaited monograph, adding a new level of scholarship in the field. My heartfelt gratitude goes to the foundation's trustees for their enthusiasm and confidence in the project.

My appreciation also goes to Garth Clark, president and director; Mark Del Vecchio, vice-president and director; and Osvaldo Da Silva, associate director, Garth Clark Gallery, New York; and Frank Lloyd, Frank Lloyd Gallery, Santa Monica. They have been steadfast supporters of Kurt Weiser's career for many years. Without their assistance in fielding queries and locating critical works, the project would not have the depth it possesses.

Many private collectors and institutions have been generous in loaning works for the exhibition. It has been a pleasure becoming acquainted with many who share a passion for the artist's work. It is difficult to loan beloved objects for over two years; and I extend my heartfelt thanks to all of you. A complete list of lenders can be found on page 90. Chip Clawson, Richard Notkin, Jacquie Rice, as well as the artist's parents, have provided meaningful insights into their relationship with Kurt. These personal reflections have deeply enhanced the project.

This exhibition and attendant publication would not have been possible without the institutional support of Arizona State University. My sincere thank you to Arizona State University President Michael Crow; Dr. Kwang-Wu Kim, dean of the Herberger College of the Arts; Marilyn A. Zeitlin, director and chief curator of the ASU Art Museum; and Senior Curator Heather Sealy Lineberry. Karyn Murphy, the Ceramics Research Center's program coordinator, was most helpful in her research and compilation of bibliographic and exhibition history as well as handling the requests for rights and reproduction. Anne Sullivan, registrar, and Laura Wenzel, assistant registrar, managed loan forms, packing, and transportation of works. In the Business Office, I wish to thank Kathleen Wacker, business manager, and Dawne Walczak, senior office specialist for their support in facilitating special events and overseeing fundraising efforts. Laura Stewart, curator of Education, was especially helpful in the coordination of educational programming related to the exhibition. Thanks also to Stephen Johnson, chief preparator and his staff, Fausto Fernandez and Chris Miller, for their expertise in packing and crating the works in the exhibition. It is my good fortune to be surrounded by such able associates.

Last, I would like to acknowledge the support of all my friends and colleagues throughout my museum career. Too numerous to mention, they have enriched my life beyond expectation.

Baby Boom Daydreams:

The Early Ceramics of Kurt Weiser

by Peter Held

Many shared experiences shaped the values and attitudes of the American generation born just after World War II. The rapid development of suburbia during the Eisenhower presidency expressed the general euphoria and abundance of economic good times. During the innocent years of the 1950s, television replaced radio as the most popular media outlet, with popular shows like *I Love Lucy, Ozzie and Harriett,* and *Leave it to Beaver* transmitting varnished glimpses into the idealized American family life. The Lone Ranger, Buck Rogers, and Flash Gordon provided heroic visions of the future; the Three Stooges, Captain Kangaroo, Howdy Doody, and Soupy Sales were also iconic cultural figures. Glandular-stimulating starlet Marilyn Monroe and hormone-pumping bodybuilder Charles Atlas branded adolescent minds, especially those of prepubescent boys. Boyish youth grew to manhood in America's heartland while reading intoxicating tales in such magazines as *Argosy* or *Boys' Life.* In many respects, Kurt Weiser, who was born in 1950 in Lansing, Michigan, was shaped by these American standards, and the socio-psycho dramas and fantasies of this era continue to play into the art he has produced for the past forty years.

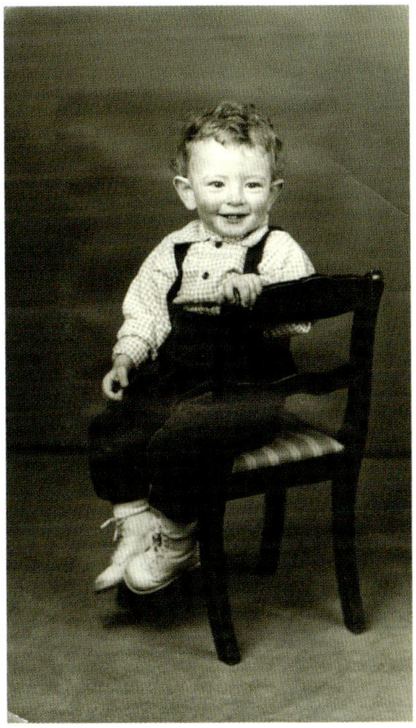

The artist, age two. He was born in Lansing, Michigan, and had one brother and three sisters.

By the 1960s the optimism communicated through the print medium and the airwaves was inflected by seismic cultural shifts. The Civil Rights Movement, the heightened nuclear threat of the Cold War, and the quagmire of the Vietnam War caused dissatisfied teens to question core values and assumptions of the status quo. Rebels Marlon Brando, James Dean, and Elvis Presley trumped Ricky Nelson and the Hardy Boys in the emerging counterculture. In the visual arts, the painterly sensuality of Abstract Expressionism, which had offered an alternative to Modernism, succumbed to Pop and Minimalism.

Kurt Weiser, the oldest of five children, was nurtured in a loving home in one of the new subdivisions that had transformed Midwestern farmland into suburbia. Weiser sought refuge and adventure in the neighboring woodlands and relished road trips to an old mining and lumber town in the Iron Mountain region of Michigan's Upper Peninsula, where his father's family lived. His mother encouraged his interests by purchasing the *World Book Encyclopedia,* which sparked his vivid imagination and offered him endless hours of reading. Kurt remembers convincing her to buy the family's first record player. "I wanted to listen to Elvis and all that kind of stuff. Later, I got my own radio. I remember the first time I heard Bob Dylan on the radio. I was amazed by the manner in which he could paint such strange and interesting pictures with his words."[1] Weiser started playing drums in his early teens, an enjoyment he still continues.

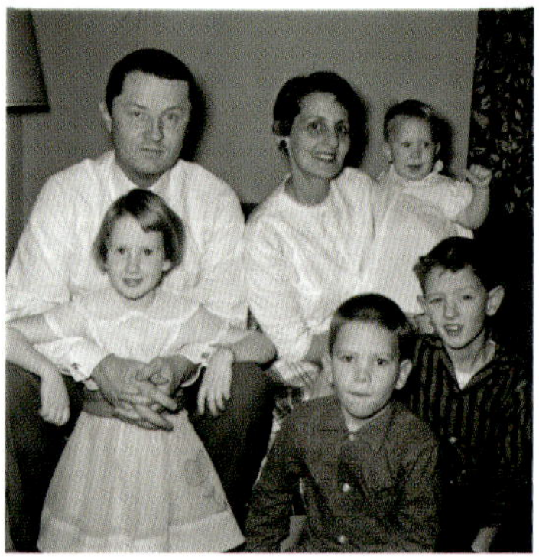

Family photo, 1958. Background, left to right: Kurt's father, Douglas, and mother, Elizabeth holding Jaime. Foreground left to right: Marybeth, Tom, and Kurt. Sister Jennifer came later.

Weiser was unhappy at the parochial school he attended in his early years. "The experience just frightened me. As a child it all seemed so dark and threatening: the Latin, the rituals; I had no basis for understanding it. With the strict disciplinary environment, I felt I was trapped in a strange and scary dream. Not much attention was given to sports or the arts; I longed to be out running with normal kids." His parents transferred Kurt to a public school near the campus of Michigan State University, which was better suited to his emerging talents. His fourth-grade teacher Mr. Harner noted that Kurt was a satisfactory student, inquisitive by nature and skilled with his hands. He was inclined to be shy and had trouble communicating, but possessed outstanding ability in art that bolstered his self-confidence. In the sixth grade, he was asked to make a tiled mural for school. Kurt Weiser, precocious yet inward, was already set on a course leading toward art and individuality.

His interest in clay developed early. Weiser first became aware of pottery at the house of a fellow student whose parents had lined their mantel with pots from their native Korea. He saw more pots on field trips to the natural history museum at nearby Michigan State University; their forms and material both struck a chord. "I remember looking at these pots and being totally fascinated. There was some sort of primitive energy that I didn't understand but it intrigued me. When I got home I tried to replicate them." [2]

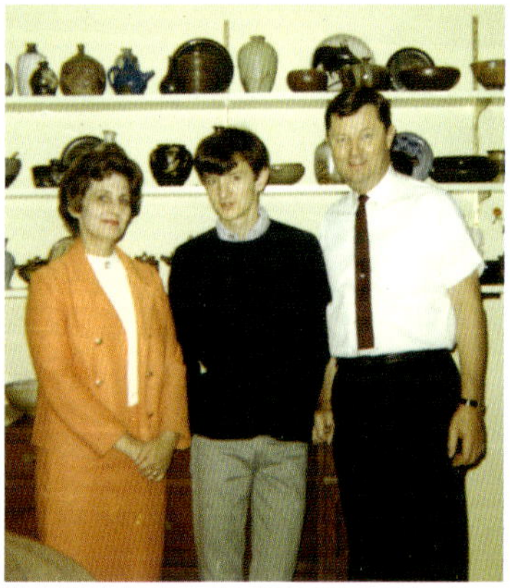

The artist with his parents, 1967. The pottery on display is from high school.

He first worked with the native clay that was churned up to grade roads and build new housing through the fields and woods around East Lansing. Advancing road graders regularly scraped up rich blue clay called Michigan slip, which Weiser characterized as "very plastic." He and his buddies used it to make "Indian" pots and pipes, which were fired in a fireplace—with mixed results. Filling the pipes with oak leaves and other tobacco-like material, they would sneak off into the woods for a smoke. Weiser refined his firing techniques with information from *Popular Mechanics*, a magazine he had checked out from the library. Enthralled by the plasticity of clay and the power of object-making, Weiser continued his explorations with the roadside clay on a potter's wheel housed in a neighbor's garage.

His interest in art was further stimulated by informal visits to M.S.U.'s Kresge Art Center and the nearby art studios. Living in close proximity to the campus, he wandered around the classrooms watching art students working in a range of media and viewed graduate exhibitions. Although he was too young to fully comprehend what he was seeing, he was attracted to this environment charged with creativity. He also spent hours at the university's library, scouring the art section and becoming fascinated with Surrealist painters and whatever else he stumbled upon.

Weiser quickly became frustrated with the "conformist expectations and endless boredom" at East Lansing High School and dropped out of the ninth grade. He took temporary jobs to keep himself occupied, from picking cherries at a nearby orchard to running a mimeograph machine in an office; he played the drums in a college rock and roll band, picking up pocket money by performing at weddings.

His parents, recognizing that Kurt had a special aptitude for the arts that would be key to engaging him in his education, drove their rebellious son up to visit the Interlochen Arts Academy (later renamed the Interlochen Center for the Arts). Located 150 miles from Lansing and originally founded as a national music camp, Interlochen had expanded its curriculum into the visual arts and become the nation's first boarding high school with a specialty in fine art. Suspi-

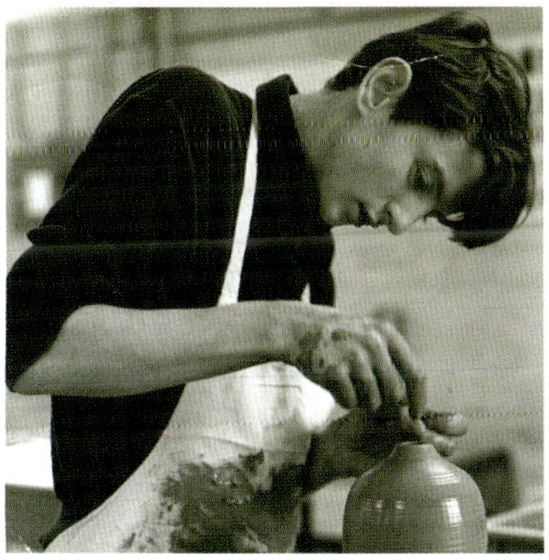

Kurt enrolled at the Interlochen Arts Academy in 1967 where he demonstrated exceptional skills.

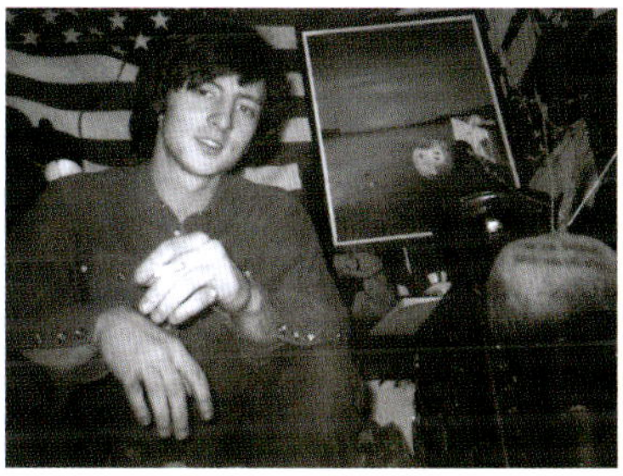

Kurt Weiser as a high school student in 1968. The Vietnam War and counter-culture lifestyles shaped Kurt Weiser's generation.

cious that this was a covert operation, that his parents were trying to trick him into going to a military school, Kurt "wasn't very impressed. It just looked oppressive and strange to me, but I knew I couldn't go back to East Lansing High School, so I enrolled. At first, the regimen and isolation proved difficult, but after I got past the uniforms and haircuts, it turned out to be a wonderful experience. The teachers were engaging and knowledgeable. It was a new world filled with interesting people and ideas, and I wanted to be part of it."

Weiser explored a strong interest in painting, stimulated especially by the work of Surrealists Salvador Dali and René Magritte. He drifted into the ceramics studio to satisfy a requirement, but like many students, he was soon seduced by the medium's tactile qualities as well as the mystery of the process. The instructor was Jean Parsons, a disciplined technician and dedicated teacher who had studied with Maija Grotell at Cranbrook Academy of Art outside of Detroit. Parsons communicated to her young students at Interlochen the Scandinavian ideals she had learned from Grotell, a Finnish immigrant who was known for her modernist sensibility. Grotell crafted utilitarian pots that were balanced, with clean symmetry, and glazed in hues and patterns of copper reds, turquoise, and browns.

Kurt was assigned responsibility for mixing clay and glazes, as well as loading and firing the kilns, to satisfy requirements of a work-study scholarship. He absorbed all that Parsons had to offer, but also deliberately challenged her artistic sensibilities. "I found a book that had Peter Voulkos's pots in it and told her I wanted to make pots like those. A teapot that had iron spots in the glaze was particularly appealing to me. She said, 'No iron spots in my class,' calling it, 'frivolous.' I looked the word up, thinking it meant dangerous, but discovered in this context it was just her opinion." By scraping the inside of a rusty garbage can, Weiser collected rust chips that he mixed it into the white clay to produce pseudo-reduction glazes, similar to the effects achieved by Voulkos. This had the added benefit of irritating Miss Parsons.

Kurt Weiser graduated from Interlochen in 1968, receiving many awards that recognized his prescient talents. These included a National Scholastic Gold Medal, seven Northern Michigan regional awards, three gold keys in state competitions, and an achievement award in ceramics and sculpture from Interlochen. A week after graduation, he had his first solo exhibition at the Lansing Community Art Gallery. He was offered a full scholarship to continue his studies at the Art Institute of Chicago, but he wasn't prepared to live in an urban setting and returned for a while to the safe haven of his parent's home.

Meeting His Mentor or Tough Love at KCAI

Weiser attended classes for a semester in the ceramics program at the Detroit Society of Arts and Crafts, but soon became disillusioned with the limitations of the facilities and, years later, recognized it was too close to home for his restless youth. Remembering a fellow student at Interlochen who had continued her studies somewhere in Kansas, he investigated art schools in that state and by a random act of providence, selected the Kansas City Art Institute, which was known for its preeminent faculty and intensive course of studies preparing undergraduate students for a solid grounding in the visual arts. Although he was not very familiar with the program, he applied and was accepted.

Weiser flourished in the creative environment of a professional art school but not without some initial adjustments. "When I first got to Kansas City, I had to take Foundations, design

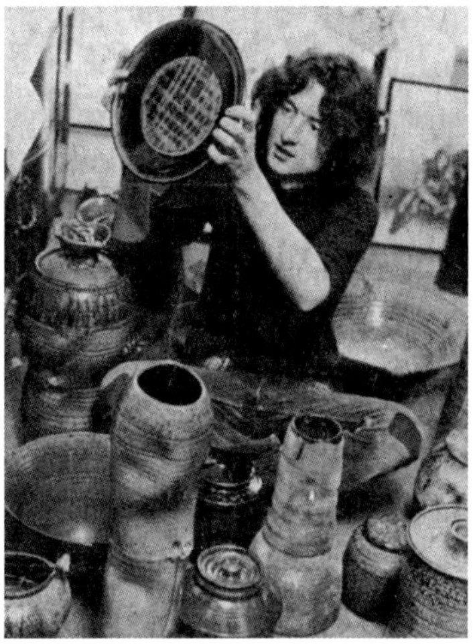

Kurt Weiser with work after completing his studies at the Kansas City Art Institute, age 22.

projects, to prepare you for deciding a major. I already knew what I wanted to do when I got there: I wanted to do ceramics and was frustrated by spending the semester doing these foolish exercises. Of course, they weren't really foolish, but I had been through this sort of thing at Interlochen and was anxious to get to work. I started sneaking over to the ceramics room, and one night I made a large group of teapots and set them out to dry. When I came back in two days, they were gone. It turned out somebody had broken them, but Ken Ferguson had seen them and wanted to know who made them. I knew nothing about him other than his fearsome reputation. I just wanted to make pots and so I took a chance and went to see him."

Ken Ferguson, an Alfred-trained ceramist, enjoyed professional success even before he had started teaching at the Art Institute in 1964, having come from the Archie Bray Foundation in Montana, where he directed the residency program for six years and gained national recognition for his functional pottery. Ferguson quickly recognized Weiser's abilities and determination, which he continued to encourage with forceful guidance and persistent mentoring. After a brief informal interview, he allowed Weiser to enroll in the ceramics program the following semester. Fellow students John Gill, Rick Hensley, Stanley Welsh, Richard Notkin, Silvie Granatelli, Donna Polseno, and Allan Winkler were talented clay artists who came of age in the 1980s. Weiser also met his future wife, Christy Lasater, who enrolled a year after Kurt, studying graphic design before switching to the ceramics program.

Richard Notkin still vividly recalls his initial encounter with Kurt in 1969. "You couldn't miss him. He wore a black leather motorcycle jacket, the kind with about a dozen diagonally zippered pockets. He had the demeanor of a young Bob Dylan, complete with long, unruly hair and dangling cigarette. But beyond this facade, Kurt soon had both the teachers and students enthralled and a bit intimidated with his phenomenal skills with clay and unbounded energy. Ferguson quickly dubbed him "Wizard." Everyone else began spending more time in the studio, trying, but failing, to keep up with Kurt, as the level of competition skyrocketed in an already highly dynamic department."[3]

Ferguson, head of the program that included Victor Babu and later Jacquie Rice, taught by example. "Ferguson fired his work at school and would discuss what he hoped he was going to find in the kiln. Every piece coming from the kiln taught us something; it was highly instructive but incredibly entertaining too. Ken had a magnetic personality and was uncanny with his ability to quench your hunger for information without spoon feeding it; he drove you to figure it out for yourself. He encouraged us to visit the Nelson-Atkins Museum to see their extensive ceramic collections, which he utilized as an integral part of the program."

Seeing competition as a life lesson, Ferguson would orchestrate scenarios, pitting students against each other. "He'd come over to me whispering confidential news. Saying, 'You should see what Hensley (or another student) is doing over there: the greatest work I have ever seen in my whole life.' So you would wait a few minutes and wander over to check it out, thinking, 'I better work harder.' Ten minutes later he would do the same thing with Rick Hensley. … His philosophy was preparing us to go out in the world. He taught us how to make glazes, how build kilns, how to do everything ourselves, be self-sufficient. Calling somebody to build a kiln for you was out of the question. It would be an embarrassment. What's most important is believing in yourself, in your work."

Setting a high standard of artistic excellence, Ferguson and other faculty members mentored students who took initiative, demonstrated a strong work ethic and promise in developing

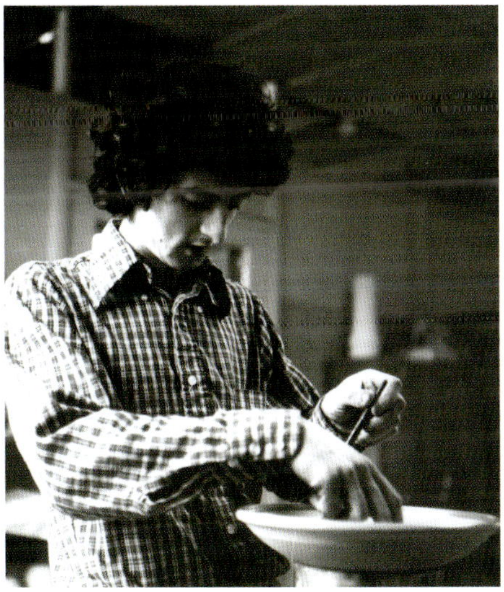

Shortly after completing his M.F.A. degree from the University of Michigan, Weiser took a teaching position at the Portland Art Museum School in Oregon, 1973.

an individual aesthetic vision. This model had served the faculty well; many of their students went on to teach at universities and colleges across the country, had successful careers in galleries, and were in demand to exhibit their work both nationally and internationally. In the catalogue for a 1983 exhibition of distinguished alumni, Ferguson reflected on Kurt's student years. "He was a little guy but what a worker! We couldn't lock him out of the building. In fact, his talent and his amazing productivity impressed the faculty but only served to alienate some of the students. They couldn't keep up with him, which frustrated many of them." [4]

The Way West

With graduation nearing, Kurt, with Ferguson's encouragement, applied for graduate school to Alfred University's School of Ceramics, where both Ferguson and Babu were alumni. Although he was accepted and was flattered with this endorsement, Weiser was uneasy with the prospect of going to a school that "every successful person in ceramics I ever knew or heard about went to. I thought it might be more advantageous to go elsewhere, I wasn't sure why, so I decided against it."

During this time, Ferguson received a phone call from potter Tom Coleman, who was calling from Oregon, looking for someone to teach ceramics at the Museum Art School of the Portland Art Museum. Ferguson turned to Kurt, who was in the office at the time, and asked if he wanted a job. "He thrust the phone so Tom could provide details of the job to me. Ken nodded his head up and down, signaling that I should accept, which I did."

Returning to Michigan after graduation, Weiser learned how to drive, bought a pickup truck, and headed west to Portland, Oregon, which had a rich ceramic tradition and at this time also had a burgeoning ceramics scene. In the 1930s potters Victoria Avakian Ross and Lydia Herrick Hodge had started of the Oregon Ceramic Studio and introduced a curriculum for adults and children. Immediately after World War II, area universities and colleges initiated ceramic programs. William Creitz, Ray Grimm, Ken Shores, and Bennett Welsh became prominent clay artists in Oregon during the 1960s, an era when hordes of counterculture potters populated the state, selling wares at craft fairs to earn a modest livelihood. Weiser taught one year at the museum art school, but it wasn't a successful fit. Coleman had some trepidation about Weiser's administrative skills, worrying about his grading of students and filing of necessary paperwork. The ceramics studio, located in the basement of the school in the heart of the city, held little appeal for Kurt's sensibilities. "I was 23 years old, and this was my first professional position teaching. Tom helped me in many ways, but our perceptions differed; I was not mature enough to notice." Soon, another position opened up at Marylhurst College where potter Patrick Horsley had been teaching; they ended up trading jobs.

During this period, Weiser focused on making a living by selling functional stoneware pots; he emulated an Asian aesthetic like many others at the time. He and fellow potter Mick Lamont would produce a truckload of pots, then drive to the Ann Arbor Street Fair in Michigan to sell them, earning sufficient funds to sustain them for the coming months. Kurt stayed for another year in Oregon, but felt restless, tired of the cycle of eking out a livelihood and yearning for a stimulating atmosphere, similar to that he had experienced at the Kansas City Art Institute. He considered returning to school and remembers "calling Jacquie Rice, who taught in the graduate program at the University of Michigan, late in the summer and asking

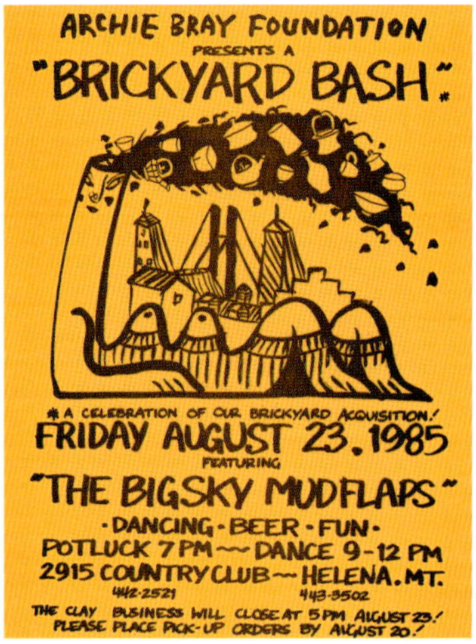

For decades, Archie Bray resident artists celebrated a busy summer with a "Brickyard Bash." Poster design by Akio Takamori.

her about the possibility of enrolling for the upcoming fall semester. I thought I was too late, but she was very supportive and got me in."

At Michigan, Kurt continued to refine his skills and spent most of his time experimenting with raku firing thrown and altered forms. "Kurt was one of those one-in-a-million students," recalls Rice. "Not only could he master any technique and make anything his heart desired, he also kept making little discoveries with materials and processes that he'd share with anyone who wanted to know. He had an energy that's both infectious and low-key—I know that's contradictory! He never gave other people the feeling he was a know-it-all, even though he was. His intellectual abilities were nearly masked by his excitement for process; now when I see his mature work, I realize that he has an intellectual depth that is almost at odds with his love of process. Yet, true to form, he puts the two to work for himself making visual poetry. A true artist can make us forget about process as it serves ideas. This is what Kurt has done."[5]

In 1976 Weiser graduated with an M.F.A. degree and found himself at another career crossroads. There was a glut of ceramic graduates searching for teaching positions at the time. Ceramists of the previous generation, many of whom had benefited from the G.I. Bill to complete their education, were in mid-career and so only a handful of positions opened up each year. Landing a tenured position was tricky. Weiser pursued one such opportunity at Kansas State University in Manhattan. After interviewing for the job, he visited Ferguson, seeking fatherly advice. "Ken didn't think this job was a very good idea. He asked me, 'Why don't you just go make pots?' But then he mentioned that the Archie Bray Foundation was looking was looking for a new Resident Director."

The Foundation, located in Helena, Montana, was founded in 1951 and sited in the defunct brickyards of the Western Clay Manufacturing Company. Archie Bray was a second-generation brick maker, a patron of the arts whose vision was to provide an unstructured and nurturing environment that stimulates creative work in clay. The Bray, through its artist residency program, never aligned itself with any strict aesthetic school or doctrine, rather drawing its strength from the generations of artists who have pursued personal artistic goals while allowing shared ideas and techniques within its borders. The Bray is self-supported, running an active clay business supplying potters and schools, offering community classes, and hosting fundraising events for an active membership.

"I had heard about the Bray for years because of Ken's involvement as Resident Director. He was always showing us slides of David Shaner's work. Shaner [who succeeded Ferguson as Resident Director at the Bray from 1963 to 1970] has always been someone I admired; I wanted to be David Shaner, starting back in my undergraduate years when I first was exposed to his work in books. Ken said, 'If you're interested, write a letter to David Shaner because he's heading the search committee.' The Bray Foundation was legendary. I remember thinking about this job and talking with Christy about the prospect of moving to Montana. As soon as I applied for the position, all those other teaching positions seemed uninteresting. I really wanted the Bray job because it was more a do-it-yourself thing. I was invited to interview and flew to Missoula and met with Rudy Autio and David Shaner. They seemed bigger than life to me, then twenty-six years old. I went home and agonized over this thing, because now I couldn't imagine doing anything else. I was really impressed with Montana: I'd never seen mountains and space like that before. Finally Shaner called back and offered me the job; I accepted. So we packed up and moved to Montana."

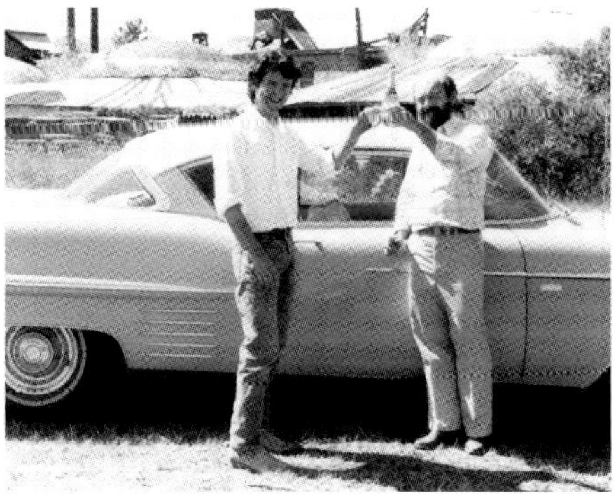

Shortly after arriving at the Archie Bray Foundation as the Resident Director, Weiser hired Chip Clawson, whom he met in graduate school, to operate the Bray's growing clay business. For years, Kurt loved driving his 1958 pink Cadillac.

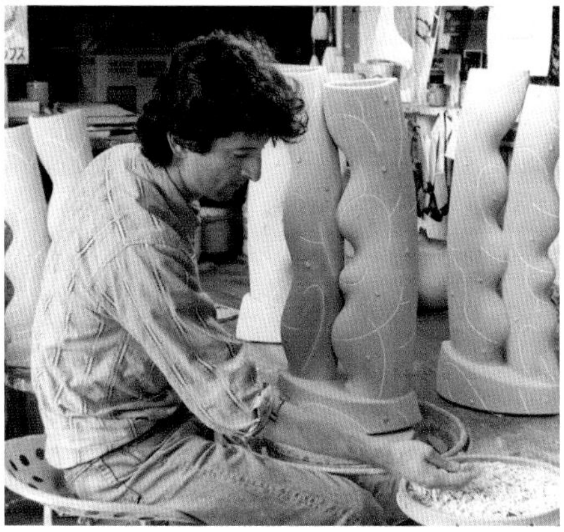

Toward the end of his Bray tenure, Kurt focused on a series of porcelain slip cast vases, glazed with a matte black or brown glaze with inscribed lines and porcelain chips, 1987.

Brickyard Bash

Like his illustrious ceramic forefathers at the Bray, including Autio, Voulkos, Ferguson, and Shaner, Weiser was in his mid-twenties when he arrived in Montana, and he really did not know what he had gotten himself into. David Cornell, the previous director (1970-1976), was poised to relocate to Oregon and ready to hand off the torch to Kurt. Weiser acknowledged the sense of urgency with some apprehension. He found the resident studios mostly empty, the glaze and clay rooms in disarray, and a hot glass shop that Cornell had built, cooled. His anticipatory vision of arriving at the temple of the clay gods had not prepared him for this reality. The next morning he drove up to Bigfork to talk with David Shaner for encouragement. Seasoned by the rollercoaster ride of running the Bray himself for seven years, Shaner had survived numerous crises. Shaner reassured Kurt, but also cautioned him that the honeymoon in the brickyard would be short-lived. Self-sacrifice was the order of the day, but the future rewards were also possible.

With $300 in the bank and winter, a brutal affair in Montana, soon to arrive, Kurt had to think fast. Realizing the fledgling clay business was a potential cash cow, he contacted area schools to pitch clay and kiln sales. His efforts were rewarded: a modicum of funds started to flow, relieving the immediate pressure. It was a DIY survival tactic. "I used to be scared to death of the gas bill, $1,800, whoa! So we had to make a lot of money."

After running the business for a year, Weiser understood that it would take more then his own energies to revitalize the clay operation. With his studio time already divided with administrative duties, he contacted Chip Clawson, who had been the ceramic lab technician at the University of Michigan when Weiser was a grad student, and invited him to visit. Remembering Clawson's love of the outdoors, Weiser routed his travel through Missoula in order to drive Clawson over the scenic McDonald Pass, showing the Montana landscape to its best advantage. His strategy worked: Clawson moved to Helena in 1977 and managed the clay business for thirty-one years. His efforts increased revenues at the Bray significantly, bringing financial stability to the organization. This strategy freed Kurt to spend more time in the studio, which boosted his production, benefiting the Bray through added revenues in their sales gallery. In reminiscing about their decades-old friendship, Clawson stated he "was always impressed by Kurt's ability to produce a large volume of consistently high-quality work, which continues to this day. He matured as a person at the Bray, coming with potential which was fulfilled in many ways."[6]

The compensation for the Resident Director of the Bray was free housing (the old chicken coop that Voulkos had remodeled for his young family in the early 1950s), a studio, and materials. By tradition half of the income from sales of work made by the Director went to the Foundation, to support the operation. In the early years of Weiser's tenure, all of the cash from sales was returned to the Foundation, out of necessity. Weiser turned out well-crafted utilitarian stoneware and porcelain plates, teapots, and casseroles. "I was trying to please everybody. Trying to please myself, trying to please the Bray, and trying to experiment a little bit at the same time. I had to keep the sales gallery stocked with work I thought people would buy. I loved it, the whole business, orchestrating it all—making the pots, getting them right, with lids that fit, glazes that look nice, firing the kiln right—the whole thing appealed to me. The process and the materials are so seductive that it can consume you."

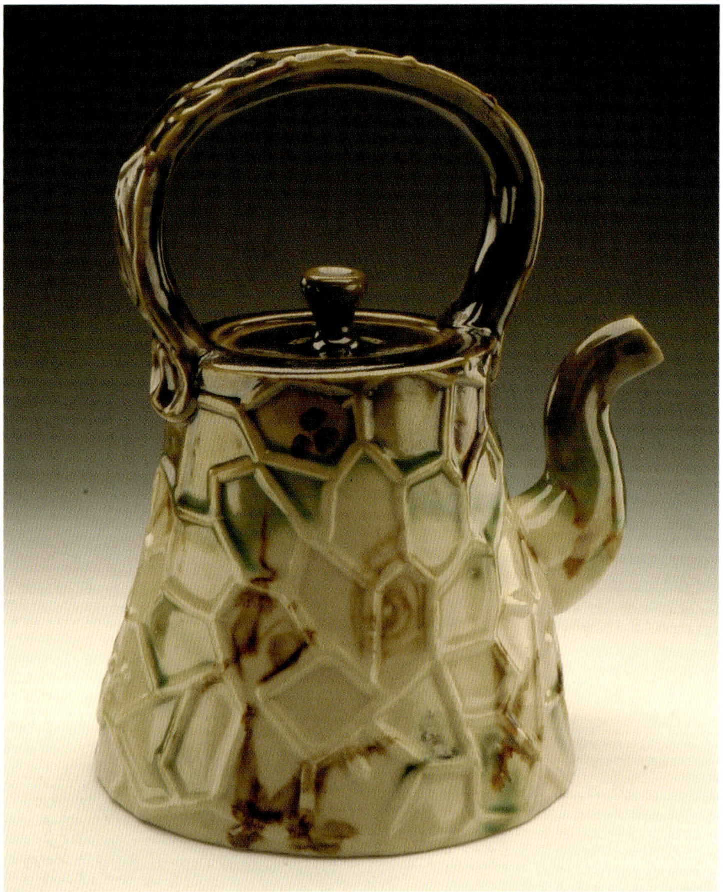

Teapot, ca. 1978
10 x 6¾ x 7½ in.
Glazed porcelain

Weiser's work from his early Bray years refined the idiom he'd worked in at the University of Michigan and at the Kansas City Art Institute. With *Teapot*, ca. 1978, the technical challenge was to slab-build an Oribe-style pot similar to one he had admired at Ferguson's house outside Kansas City. A facile hand and the first hint of the surrounding natural forms seeping into his aesthetic are evident in *Plate*, 1977 and 1978. Baptized into Orientalism by Ferguson, Weiser incorporated slip designs, Shino and ash glazes, and modest forms to master his craft.

Weiser jokingly concludes that the plates and other functional ware from this period largely reflect his effort to make David Shaner pots without being Shaner. The irony, he said, is they came at a time when most other young potters were trying to make Peter Voulkos pots. By the mid-1970s, Voulkos's freewheeling style had trumped all others. His heavy forms, slashed and seared with punch-outs, fascinated Weiser. "I was seduced by the beauty and power of it," he recalled. "But I also remember feeling it was a bit of an aesthetic tar baby too, because I just saw so many people stuck to it…I knew I had to go another way."[7]

Weiser tapped past experiences to recruit new resident artists for the Bray. When he first arrived at the Bray, there were only three residents in place: Wally Bivins, Richard Swanson, and Lauren Kearns, just arriving from Seattle. Typically, artists during this era spent one to two years working independently, honing their skills while gaining knowledge about setting up a studio and operating a small business. Residents were required to mix clay once a week to assist the clay business, exhibit and sell their work through the Bray sales gallery, and participate in exhibitions showcasing the residency program at museums and galleries nationwide. In time, Kurt was able to recruit more artists, bolstering the residency program. The roster of 106 artists whom Weiser attracted during his tenure includes many alumni of Alfred University and the Kansas City Art Institute. John and Andrea Gill, Chris Gustin, Clary Illian, Richard Notkin, Bobby Silverman, Judith Solomon, Akio Takamori, David Regan, and Arnie Zimmerman worked

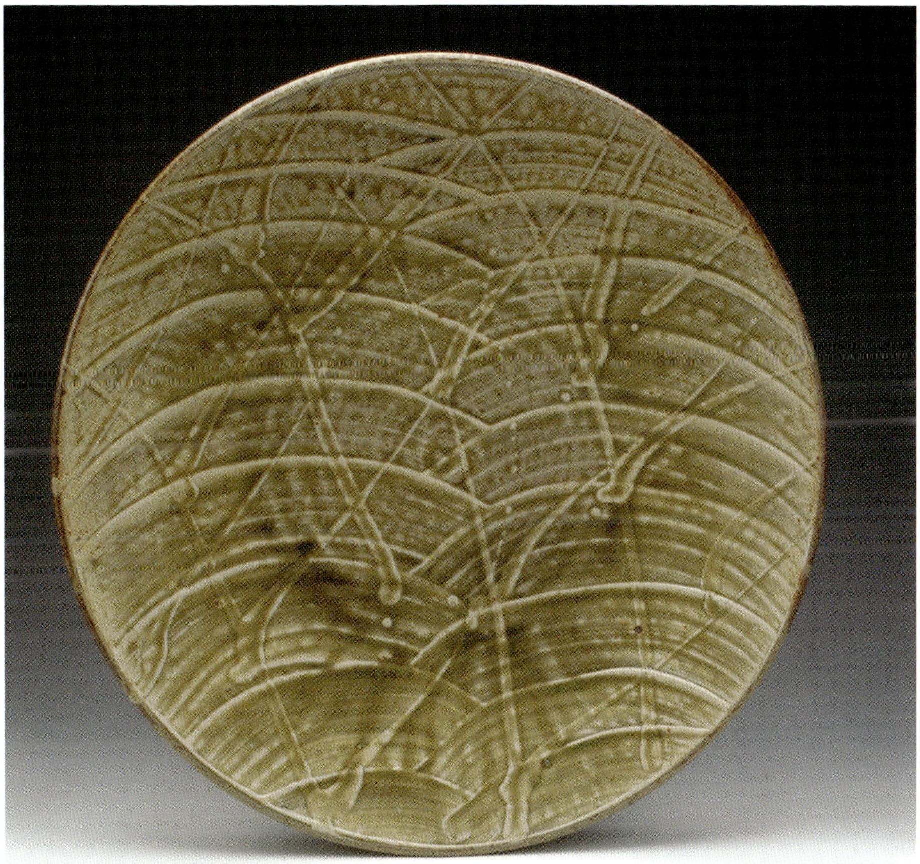

Plate, 1977
2½ × 20 in. diameter
Glazed stoneware

Plate, 1978
2½ × 21¼ in. diameter
Glazed stoneware

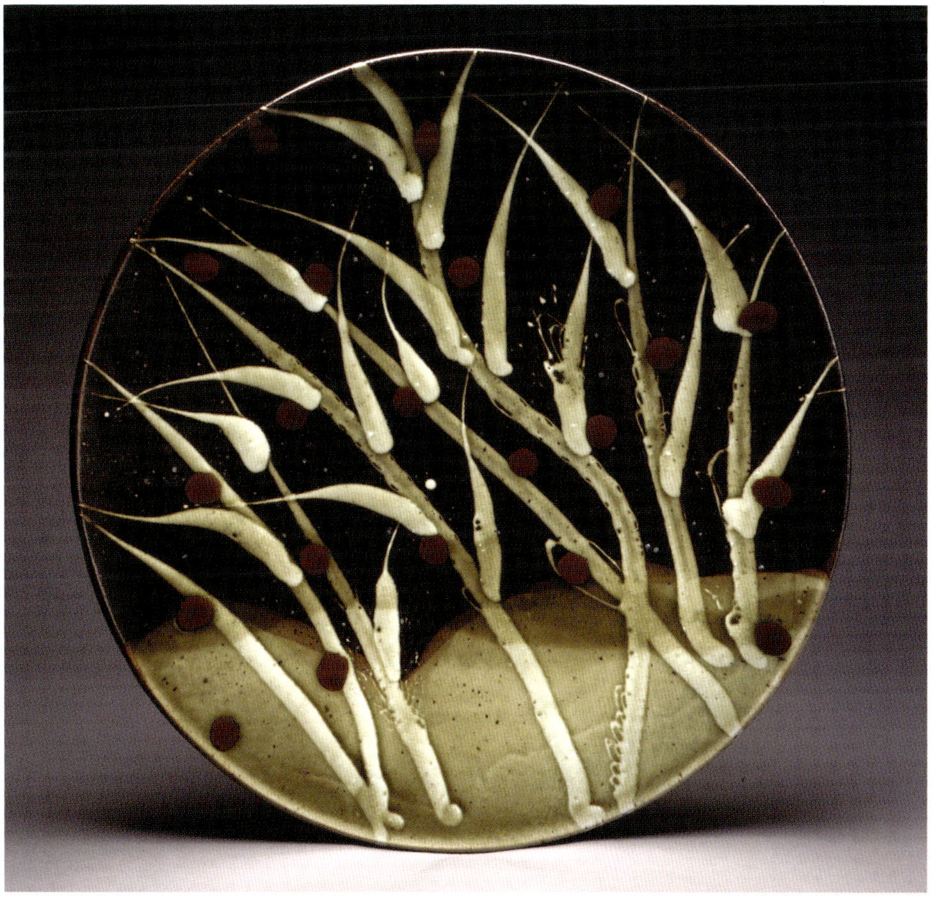

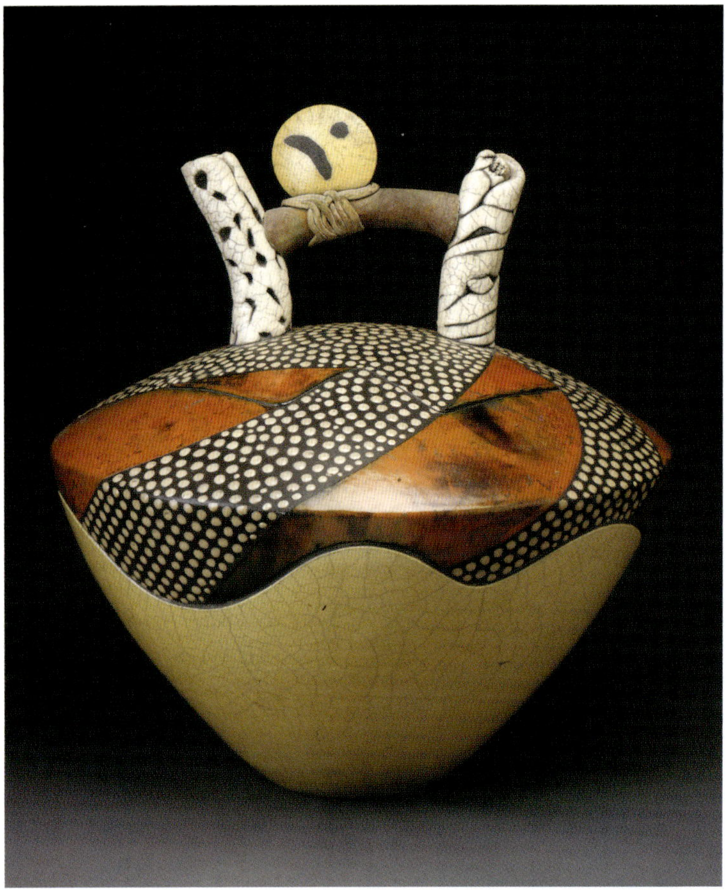

Raku Stirrup Jar, 1981
20 x 16½ x 16½ in.
Glazed earthenware

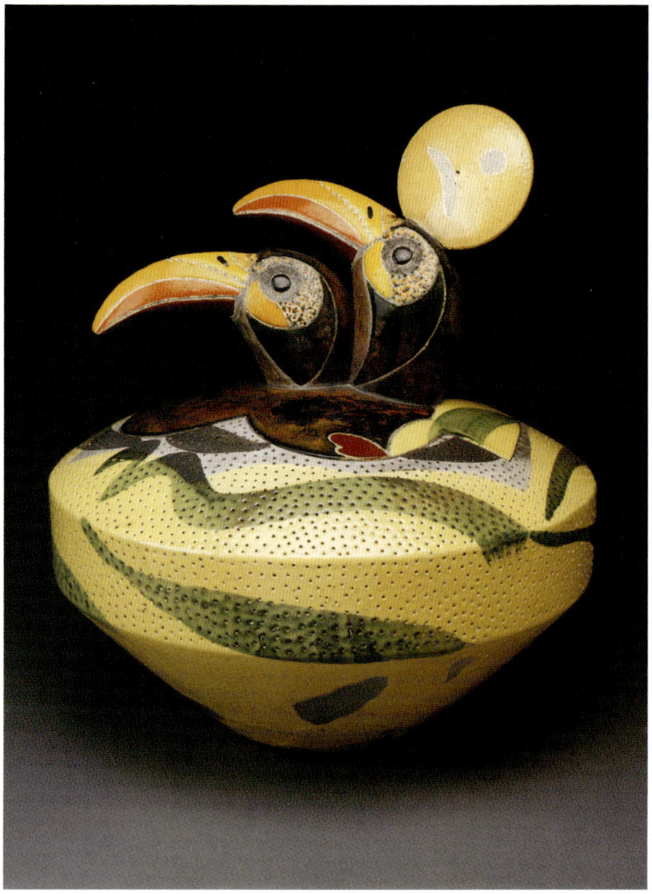

Raku Toucan Jar, 1981
15 x 12½ x 12½
Glazed earthenware

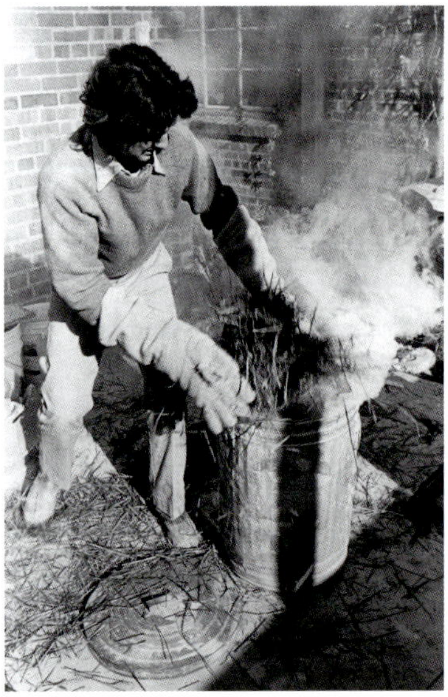

The artist continued his fascination with the raku process while working at the foundation.

in Montana through Weiser's efforts; he also attracted two future Bray Directors, Carol Roorbach (1989–1992) and Josh DeWeese (1992–2006).

Weiser gave globalism a jumpstart by expanding the Bray's program to include artists who worked outside U.S. borders during the early 1980s. Through Akio Takamori's contacts and Weiser's travels, Japanese artists Kazz and Tomoko Nakamoto, Michio Sugiyama, and Ei (Sano) Yamamoto brought refreshing new ideas to the Bray. Pang-Ling Chou, a talented Chinese artist, created inventive mixed-media teapots. Through a chance meeting with Weiser in Calgary, Canada, Suwanee Natewong from Thailand came to the Bray and created a large-scale mural that is now permanently installed in the recently opened Ann & David Shaner Resident building. This tradition has been continued by those who have followed Weiser in directing the Bray, creating an international dialogue that has enriched the resident artists' experiences.

Weiser sought to achieve the technical command needed to make the raku glazes do precisely what he wanted, a control he had never been able to accomplish at Michigan while a student. *Raku Stirrup Jar*, 1981 and *Raku Toucan Jar*, 1981, continued his exploration of and affinity toward historic Peruvian models. When most potters were drawn to the spontaneity and accidental qualities of *raku*, Weiser fashioned a more formal, sophisticated approach to the work, both in form and surface decoration, not unlike Wayne Higby's iconic landscape forms from the 1970s. Softened by the fire's contact, *Stirrup Jar*, topped by a calligraphic orb, has a mottled color palette, irregular crackle patterns, and swirling design elements.

Vessel with Stirrup Handle, 1984, was inspired by the patterns of the flashed salt-glazed brick that are produced in the brickyard when irregular areas of the brick are exposed to atmospheric conditions. Weiser also mimics the fire-cord, or *hidasuki*, effects that potters in Bizen,

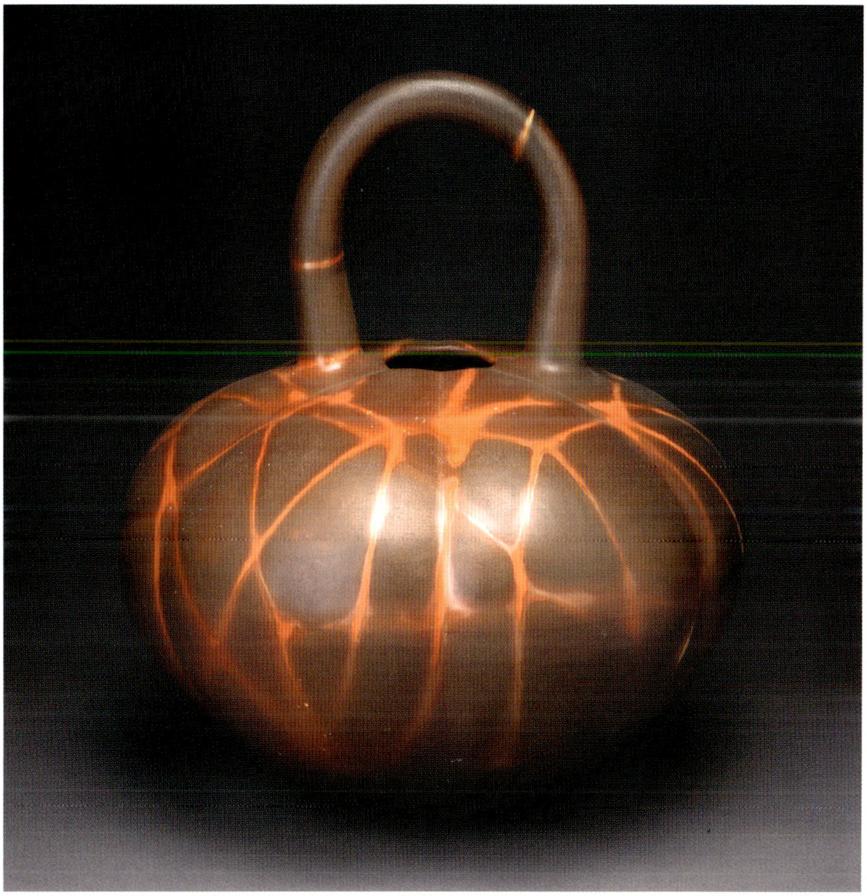

Vessel with Stirrup Handle, 1984
17 x 16 x 16 in.
Earthenware, residual salt-fired

Blue and White Crackle Bowl, 1986
5¾ x19¾ in. diameter
Glazed stoneware

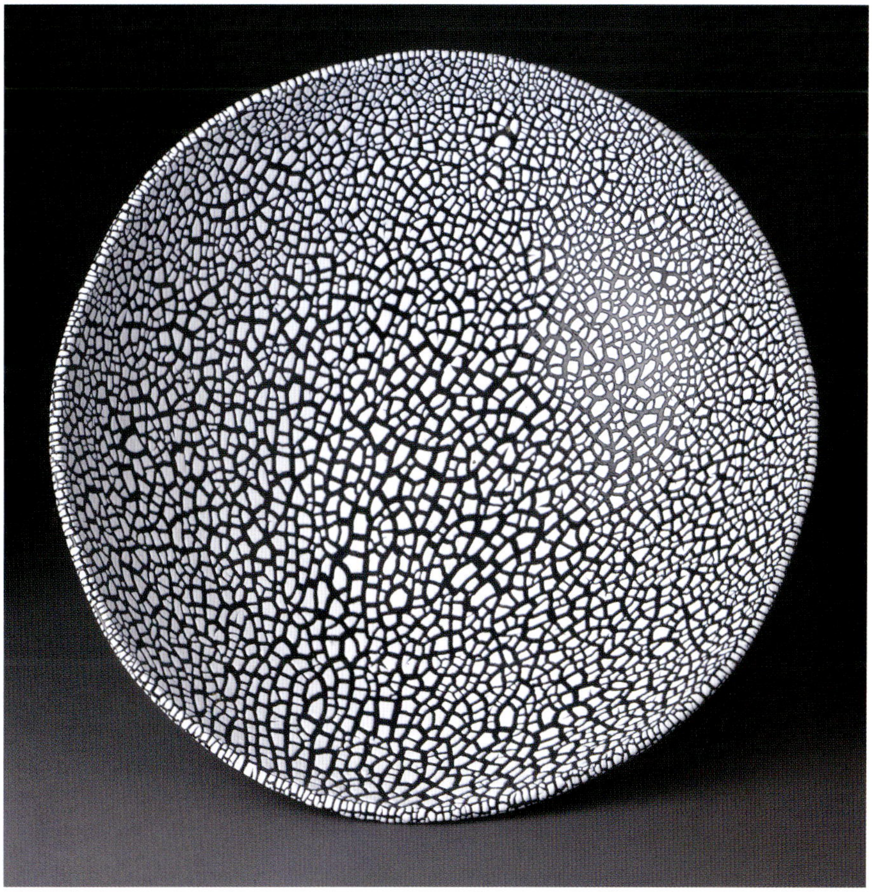

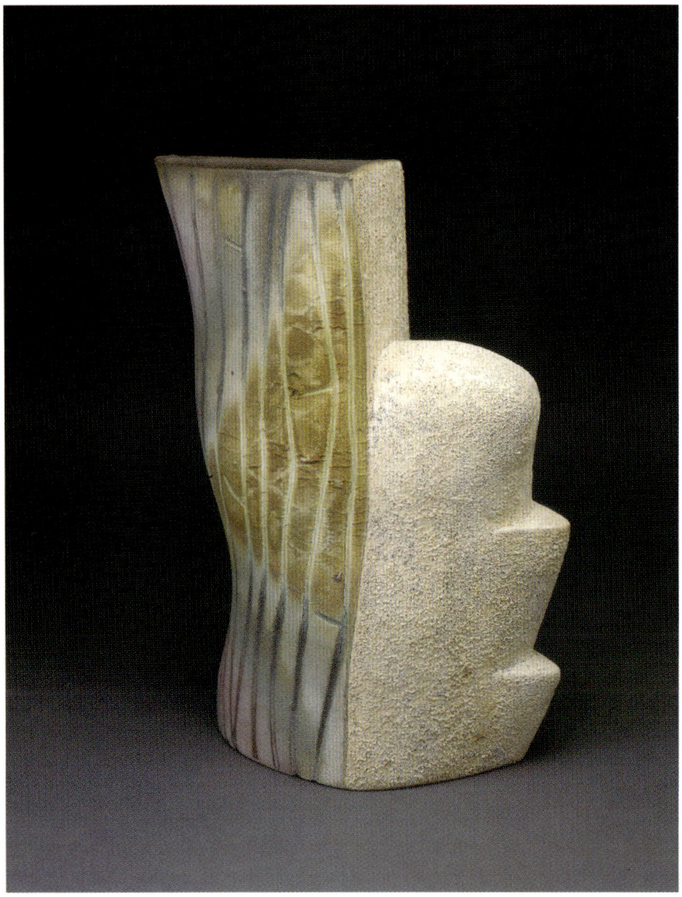

Stepped Vase, c. 1985
16½ x 9 x 5½ in.
Glazed porcelain

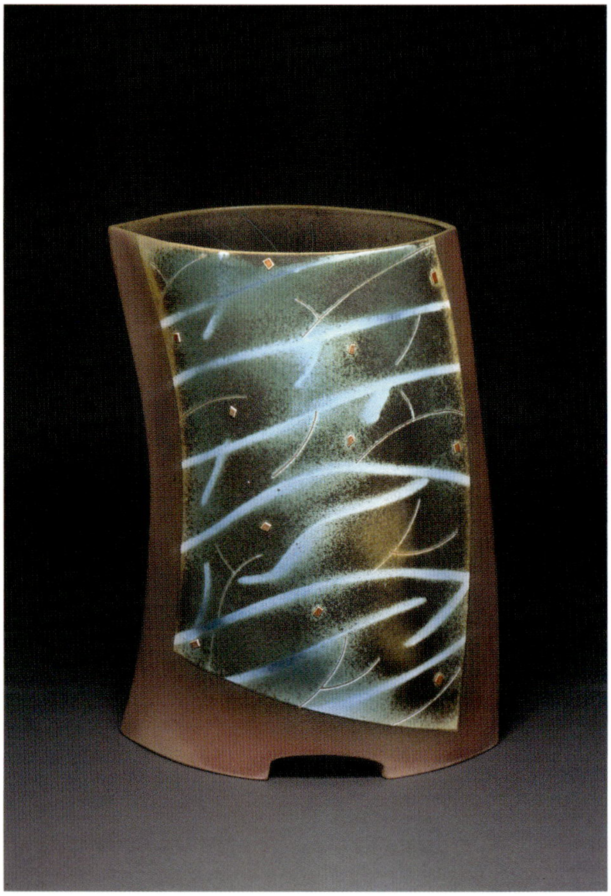

Untitled Vase, 1984
15 x 9½ x 5 in.
Glazed porcelain

on Japan's Inland Sea, produce by the traditional method of wrapping rice straw around the surface of wares before firing. Weiser engaged modern technology, using a new ceramic product (kaolin fiber rope) to create a webbed matrix that, when lightly salt-fired, reveals a resist pattern similar to the traditional effects of the Japanese ware. *Blue and White Crackle Bowl,* 1986, demonstrates his mastery of glaze development. This cratered surface, created by the addition of magnesium carbonate, causes the applied glaze to shrink away from the clay, breaking up into islands against the backdrop of black clay.

Kurt was actively exhibiting his work nationwide when in 1982 he was introduced to the influential ceramics dealer and historian Garth Clark, who offered to mount a solo show at his new gallery in Los Angeles. Clark offered advice and guidance on building a career, a topic the artist hadn't previously considered in depth. Weiser gratefully accepted Clark's perspective and expertise, and the two built a relationship that continues to this day. "Working with Kurt is fascinating. He is an enigma—very smart, very intuitive, and yet opaque. One gets glimpses into his mind, nothing more. I am never sure what he is thinking, but he is a pretension-free zone. I just accept what arrives."[8]

A major change swept the Bray in 1984 when the brickyard adjoining the pottery, which had been part of the Bray property until 1966 when it was sold in a foreclosure auction to satisfy debts, was acquired for $75,000 from the Medicine Hat Brick and Tile company. Kurt, along with Chip Clawson, William Porter, and David Shaner, negotiated with Gordon Sissons, owner of Medicine Hat Brick and Tile, who at that time also owned the brickyard property. Weiser credits the business acumen and influence of Bray board president William Porter with community leaders for the successful conclusion to the negotiations. The acquisition of the 13 acres and historic buildings of the brickyard, lovingly referred to as the "Louisiana Purchase," allowed

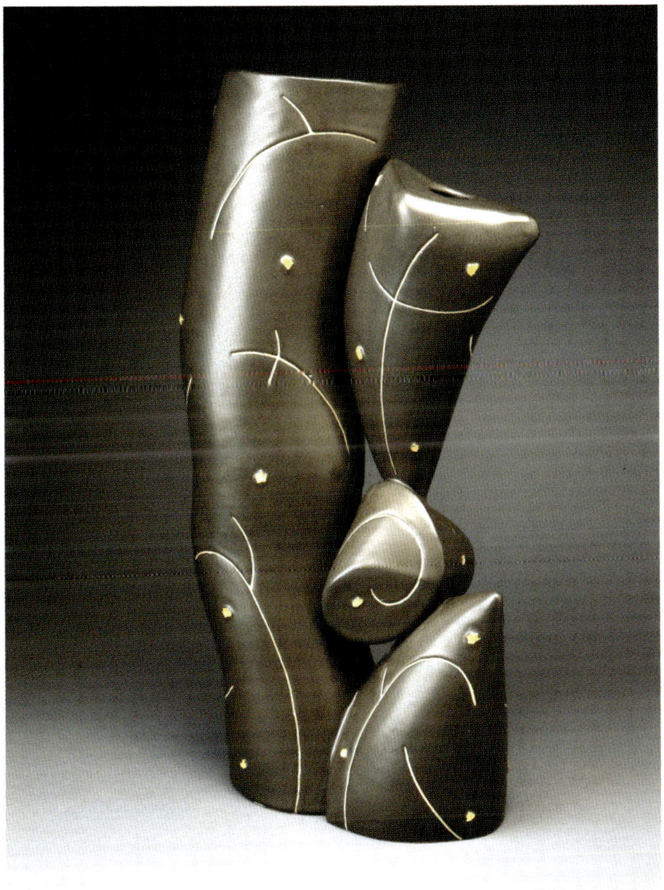

Tango, 1988
22 x 10 x 5 in.
Glazed porcelain with porcelain inclusions

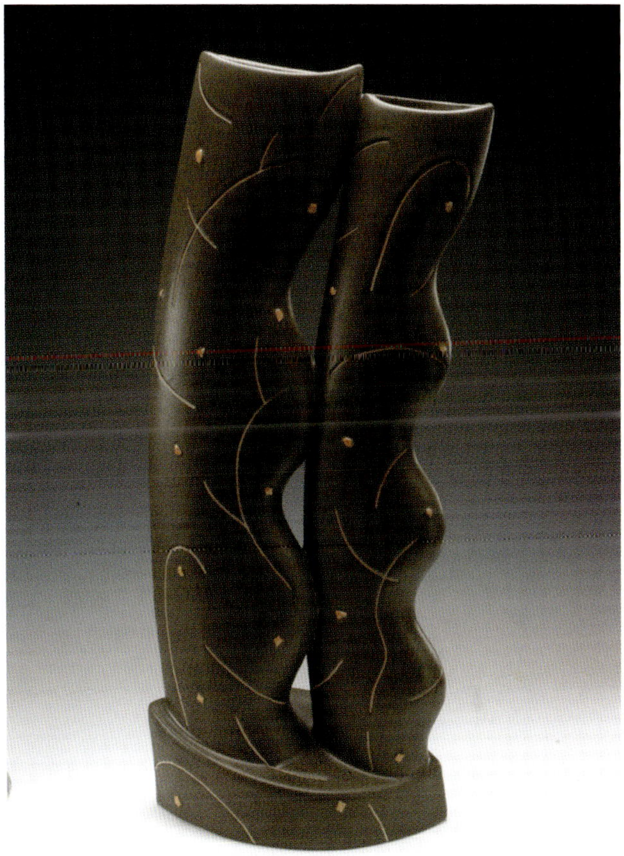

Vessel, 1988
25 ½ x 10 ½ x 5 in.
Glazed porcelain with porcelain inclusions

Weiser to expand both the residency program and clay business.[9] Prior to the purchase, there was only studio space for a half-dozen artists; this transaction allowed for the residency program to double in size.

Migrating out to the brickyards, artists started to explore the raw material left behind, creating site-specific works for the first time. Robert Harrison, who was a current resident, leapt at the opportunity to stretch himself artistically. He created three architectural works in the next few years: *Tile-X* (1985), the *Potter's Shrine* (1985–87), and *Aruina* (1988). This opportunity became a springboard for Harrison's career, as he continued to produce work in this vein throughout the world over the next two decades.

Kurt's output toward the end of his Bray tenure focused primarily on slip-cast porcelain forms. *Stepped Vase,* ca. 1985, and *Untitled Vase*, 1984, reflect a growing restlessness with producing a high volume of functional pottery and the desire to explore new concepts. He experimented with spraying barium-rich glazes on cast porcelain forms, which produced saturated colors and surfaces embellished with a patina of antiquity. These scarified and glacier-worn pieces, which seem to have emerged from an archaeological context, attest to the artist's love of surface and pattern and mirror his ongoing fascination with the natural world.

The last series that Weiser produced at the Bray, which included *Tango,* 1988, and *Vessel*, 1988, are refined in appearance, largely monochromatic, with a matte black or brown glaze and sgraffito marks on the surface. *Tango* records the movement of a dancer in abstract terms, as the title implies. Graceful yet complex with undulating forms, it appears both modern and historic, drawing on Weiser's appreciation of past cultures, while expressing a

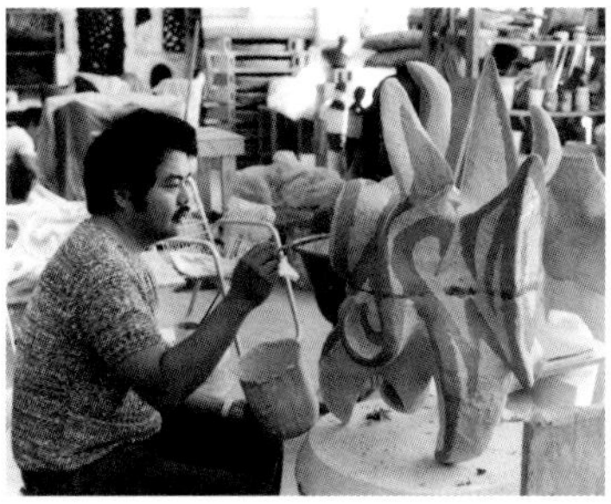

Michio Sugiyama, a Japanese artist, worked on brightly glazed abstracted sculpture during his residency.

contemporary aesthetic. Off-kilter, the works from this period are frozen in mid-pose, lending an uneasy yet elegant posture to the vessel.

Having successfully conquered the technical challenges of the series and exhausted variations of form, Weiser felt bored. Working tirelessly for eleven years, he had advanced the Bray's artistic vision and provided financial security while gaining wider recognition for his studio work. While pondering his next artistic move, he was offered a job teaching at Arizona State University, a challenge that appealed to him. In 1989, Weiser moved with his family to Tempe and embarked on the next stage of his career. Not one to rest on past success, he looked forward to new experiences and the opportunity they would present to grow once again.

NOTES

1. Unless otherwise noted, all quotes are from an interview conducted by the author with Kurt Weiser on May 22, 2006, for the Smithsonian Institution Archives of American Art, Nanette L. Laitman Documentation Project for Craft and Decorative Arts in America.

2. Interview conducted by Edward Lebow and Kurt Weiser, 1993.

3. Email to the author on December 5, 2006.

4. *Ceramic Artists: Distinguished Alumni of the Kansas City Art Institute.* Kansas City, Missouri: Charlotte Crosby Kemper Gallery, 1983.

5. Email to the author on June 19, 2006.

6. Phone conversation with the author on November 7, 2006.

7. Interview conducted by Edward Lebow with the artist in June 2006.

8. Email to the author on November 27, 2006.

9. For a more complete history of the Archie Bray Foundation, see *A Ceramic Continuum: Fifty Years of the Archie Bray Influence.* Seattle and Helena, MT: University of Washington Press and the Holter Museum of Art, 2001.

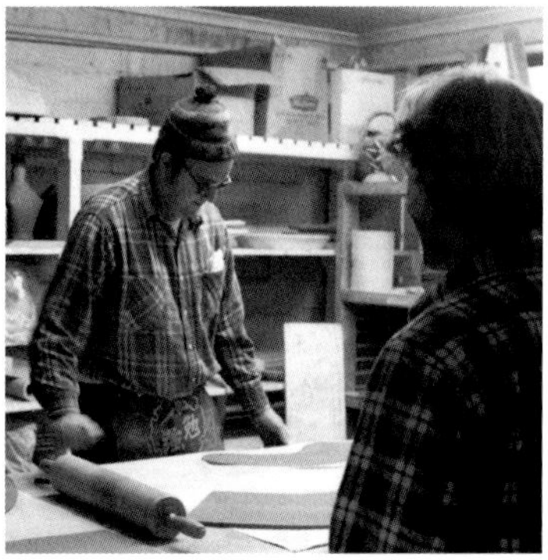

When he arrived at the Bray, Weiser recruited many former classmates from the Kansas City Art Institute. John Gill is shown working in the main pottery building. Ken Sedberry looks on, 1979.

Abandoning Gentility:
The Roots of Kurt Weiser's China Painting

Ulysses Grant Dietz

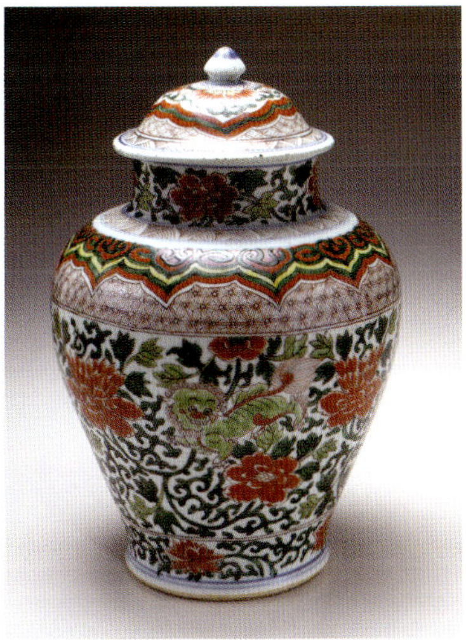

Covered jar with enamel decoration, ca. 1625, China. Height 14½ inches. Collection of the Newark Museum, Gift of Herman A. E. and Paul C. Jaehne, 1941 41.1050a,b

Kurt Weiser's surreal, lushly painted porcelains update the great tradition of American china-painting that flourished in the early twentieth century. In the years around 1900, male decorators dominated the professional side of china painting in the United States, carrying on a tradition that dates back to the birth of the American porcelain industry.[1] Edward Lycett, seen as the father of American china painting, was trained in England's ceramic industry. After emigrating to the United States in 1861, he became a traveling prophet of porcelain decoration here.[2] It was his influence that, by 1900, had helped create a vast body of amateur female china painters across America.[3] In contrast, as the American porcelain industry grew, particularly in Trenton, New Jersey, it was the continuing practice for manufacturers to import male decorators from Royal Worcester and from other factories whose work they wished to emulate.[4] The resurgence since 1975 of the decorative surface in contemporary ceramics has become not only acceptable but an important restatement of past traditions. In the transition from Weiser's earliest works in this exhibition, which embody the Shoji Hamada-Bernard Leach aesthetic of earthy, Japanese elegance, to his current painted pots and enameled vessels, one can see the reemergence of decoration in post-modern ceramics unfold in the career of a single artist.

The Chinese began making what is known today as porcelain during the T'ang Dynasty in A.D. 618. They discovered that combinations of kaolin clay and feldspar resulted in a clay recognized for its brilliance, translucency, and durability. China paints, or overglaze enamels, are minerals or colorants that are formulated, fired in a crucible, then left to cool. They are ground up with an oil base for easier brush application and fired at lower temperature, generally over a white glaze or ceramic body. These overglaze colors are vivid in hue, and when applied on a white ground produce vibrant effects that are not possible to achieve with earth-toned clay bodies. Although most famous for its blue and white porcelains, the Ming dynasty (1368–1644) was also the period when the use of brilliantly colored overglaze enamels was first perfected in China. It was during the early Qing dynasty, during the reigns of the Kangxi and Chienlung emperors, that skilled china painting with enamels reached its zenith.[5] During this period the secret to porcelain production was finally discovered in Europe, and with it the tradition of overglaze enameling became central to luxury ceramics produced in the West.

Kurt Weiser has a deep appreciation for Ming Dynasty blue-and-white wares, especially the contrasting cobalt underglaze on pure white porcelain (the four cups and bottle, catalogue Pl. 19 and Pl. 21 demonstrate this influence). However, the later Ming and Qing enameled wares are evoked most vividly by his current work. This lidded temple jar, from the late Ming

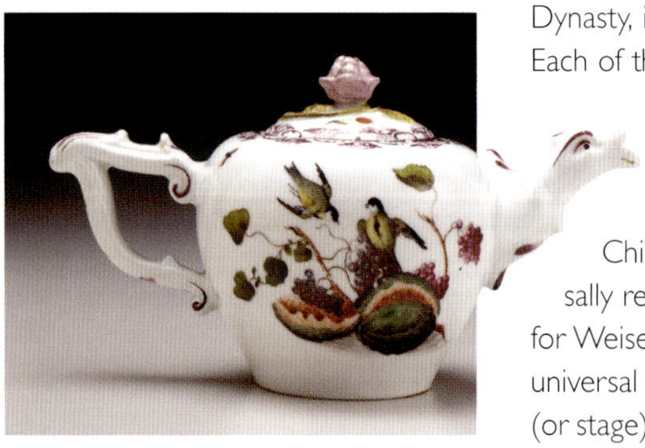

Teapot, ca. 1750, Royal Vienna Porcelain Works. Height 3¼ inches. Collection of the Newark Museum, Gift of Sarah W. Downer, 1914 14.701a,b

Dynasty, is covered in a dense tapestry of red, green, and yellow enamel outlined in black. Each of the decorative motifs—from the peonies to the smiling, whimsical lions—had a specific cultural significance to the anonymous Chinese porcelain decorator, although no particular narrative was intended by the composition. The surreal imagery with which Weiser decorates his jars is intended to convey meaning, even when the form is conventional.

Chinese ceramics has also added to Weiser's vocabulary of forms. The ginger jar, universally recognized as the epitome of classical ceramic vessels, is used as a point of departure for Weiser, who models lopsided double-vessels and jars with asymmetrical shoulders. "This universal form seduces the viewer to look closer," the artist explains. "The pot is the theater (or stage) and the painting is the play."[6] The altered forms of some Weiser vessels gives them a vaguely mutant quality. Weiser has subverted the symmetrical, comfortable vessel form, known in the West as a "ginger jar," and given it an asymmetrical organic shape in such pieces as those on pages 1 and 62–63.

When East-West trade was established by the Portuguese beginning in the fifteenth century, Chinese porcelain of the contemporaneous Ming dynasty was the luxury good most valued by the aristocracy of Europe. It was an obsession that led to the discovery of the formula for true porcelain—white and translucent—under Augustus the Strong, Elector of Saxony, in the early years of the eighteenth century.[7] By the middle of the eighteenth century, enameled porcelains from all over Europe competed with Chinese wares for the luxury market.

A diminutive Viennese teapot from the mid-eighteenth century exemplifies this work: it is covered with motifs—birds, insects, broken fruit—that could be found on work Weiser is doing today. All of its surfaces are decorated, but in the traditional hierarchy of placement, a simplified scene, here composed of grapes only, appears on the "back" of the teapot. Unlike Weiser's compositions, there is no narrative theme, just decoration. Weiser draws on an asymmetrical naturalism that permeated the decorative arts in mid-eighteenth-century Europe—the naturalistic finial is a good example of an element that could also be incorporated into Weiser's work—but he has dramatically altered the scale and form of the vessel itself. By enveloping the entire jar with enamel and thus linking the two sides into one narrative, he has expanded both the meaning and the scale of his painting. The narratives in Weiser's pots are dreamlike (or, depending on your viewpoint, nightmarish), but his china painting could not be described as merely decorative.

The imagery in rococo china painting, often drawn from popular prints, was meant to be skilled and charming, nothing more. This limitation of content in china painting carried on throughout the nineteenth century: painted floral decoration and generic landscapes continued as standard fare on porcelains and light-bodied earthenwares. Of course, there are notable exceptions: historical figures and specific topographic scenes appear on American porcelain and appear on costly English ceramics.[8] China painting emerged in the third quarter of the nineteenth century as a profession for men and as a genteel hobby for women, buoyed by the aesthetic reform movement, which gave rise to the arts and crafts style—both of which have strong English roots. Even in the hands of professional china painters, such as Edward Lycett or William Morley, the result of skilled china decoration was, generally speaking, aesthetic and not intellectual in its purpose. The overriding decorative intent of china painting made it ideal for amateur copyists who focused on developing technical skills rather than trying to invest their work with meaning.

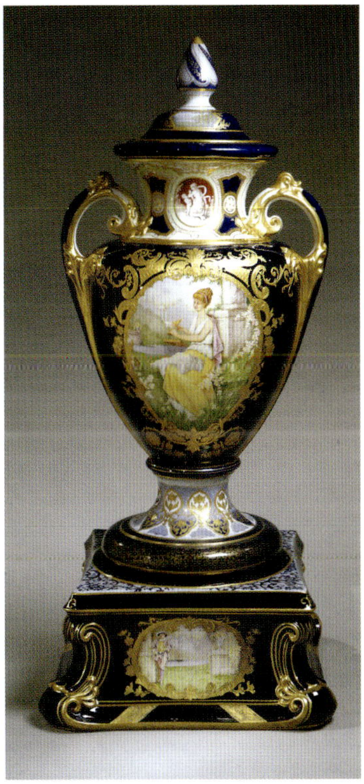

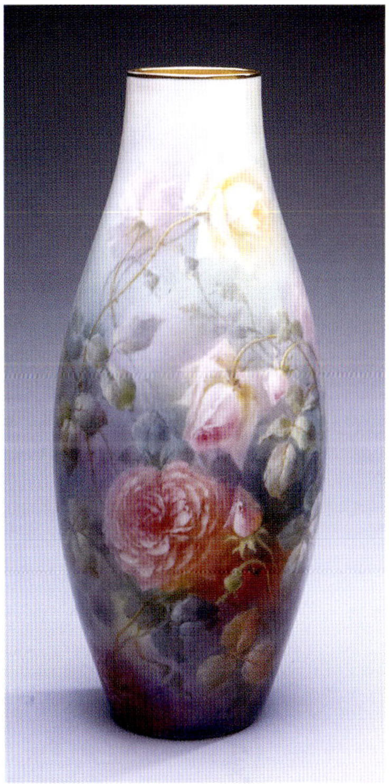

The Grecian Vase, 1904, decorated by Lucien Boullemier, Trenton, New Jersey. Height 55 ½ inches. Collection of the Newark Museum, Gift of the Crane Company, 1969 69.133a–c

Vase with Roses, 1903–1906, William H. Morley, Trenton, New Jersey. Height 16 inches. Collection of the Newark Museum, Gift of Dr. Harold H. Kelsey in memory of Arthur Colletti, 2002.38.12

In the late nineteenth century, as prosperous women took up ceramic decoration for amusement, social reformers were looking at it as a potential way to humanize women's work in an industrial age. Fixed notions of gender roles in America made ceramic decoration an ideal profession for working-class women (and genteel women in need of financial assistance), because it was not as brutalizing as factory work was. In most cases, professional female ceramic decorators (not all were china painters) were supervised by men.

In New Jersey, the center of the American porcelain industry, china painting flourished at the Ceramic Art Company, founded by Walter Scott Lenox in 1889 in Trenton. Among the most talented male china painters who worked there was Lucien Boullemier, who was born in Stoke-on-Trent, England, in 1877 and studied with his father, Antonin Boullemier, a ceramist and portrait painter working for Minton in Staffordshire. Boullemier moved to New Jersey between 1903 and 1905 and found work at Ceramic Art Company (later renamed Lenox China). Boullemier's monumental *Grecian Vase* of 1904, one of four vases he decorated for the New Jersey Pavilion at the Louisiana Purchase Exposition in St. Louis, combines technical skill and emotional heft. The idyllic landscapes and conventionalized beauties of *Grecian Vase*[9] leave no question that these were intended as works of art. Boullemier had to work within the confines of an academic art tradition that had not yet been modernized; he also had to create his composition on a form designed by someone else, as did all china painters. The shape of this piece, with its pedestal, handles, and lid conforms to traditional "palace wares" produced at continental porcelain factories in the eighteenth and nineteenth centuries. It is a self-conscious reference, both in form and in decoration, to its European counterparts, undoubtedly aiming to prove that anything the Old World could do New Jersey could do just as well. Weiser, on the other hand, because of his training as a potter, designs both the form and creates a composition to suit his "porcelain canvases."

Even more celebrated in his day was William H. Morley (1865–1935), an English china painter who decorated porcelains from the 1890s into the 1930s for the Ceramic Art Company (called Lenox China by 1906). Morley's tall vase, with lush rose blossoms spiraling around the simple vessel form, could hardly be more different from Boullemeier's work, but his lush blossoms suggest something of the supercharged landscapes in Weiser's vessels. Morley makes use of the entire body of the vessel as a canvas, anticipating the dense, closely viewed flora on Weiser's vessels.[10] Like its eighteenth-century counterparts, however, Morley's china painting was not expected to convey meaning but was limited to aesthetic and technical brilliance. The forms Morley chose to decorate were produced by Lenox, and he had no influence over their design. In this way his work would have been parallel to the enamel artists of the Qing dynasty at the porcelain factories in Jingdezhen, China, who adapted their ideas to preordained shapes.

In his surface decoration Weiser draws on the whole evolution of modern art, and is among the few brave ceramic artists who have brought emotional impact, and thus the potential for meaning, back into decorated porcelain. Contemporary potters such as Susan Thayer and the late Anne Kraus have embraced the feminine aspect of china painting in their own work. California ceramic artists Ralph Bacerra and Adrian Saxe have developed lush overglaze decoration in their work; Bacerra was influenced by Asian traditions while Saxe's elegant Antelope Jars and other forms derived from European court porcelains.

Infusing his china-painting with the influence of modern art, while paying homage to the technical skills of his china-painting forebears, makes Weiser's work appealing—and at times satisfyingly disturbing. The obvious links, both in form and in decoration, to earlier decorative arts traditions create an immediate response in the viewer—even in an uninitiated viewer. The evident virtuosity of Weiser's painting is captivating even before it is examined closely. It is on second thought that Weiser's china-painted vessels become provocative and even unnerving. As the potential darkness behind the superficial beauty begins to register in the viewer's mind, a wide range of possible interpretations is revealed, offering an experience unique in the realm of modern ceramic art.

Notes

1. *See* Alice Cooney Frelinghuysen, *American Porcelain, 1770–1920*. New York: Metropolitan Museum of Art, 1989, page 19. At the Tucker porcelain works in Philadelphia in 1828, Thomas Tucker, younger brother of one of the firm's principals, was engaged as a china painter because of his special skills.

2. *Ibid,* page 234, and note 1.

3. *Ibid,* page 242, and note 1. This is not to say that there were no professional women, Adelaide Robineau being a primary and important example. But her early interest in china painting shifted as she became a pioneer studio potter.

4. *Ibid,* page 232.

5. See *2000 Years of Chinese Ceramics, Newark Museum Quarterly*, Summer/Fall 1977, page 32.

6. Quote from Kurt Weiser, interview with Peter Held, June 17, 2007.

7. The best telling of this story is Janet Gleeson's *The Arcanum*. New York: Warner Books, 1998.

8. Frelinghuysen 1989, pages 98–99, and 102. The Newark Museum owns a large dessert service from the 1830s by the Derby Porcelain Works, each piece decorated with specific views of the Derby countryside (accession number 76.29).

9. See Frelinghuysen 1989, page 248, for a discussion of the *Trenton Vase,* and page 246 for another example by Boullemier. The three other vases were called the *Trenton Vase,* the *Woodland Vase,* and the *Rose Vase.*

10. See Ibid, pages 254–55, 256–57, for discussions of Morley's work for Lenox.

11. See Ibid, pages 250–51, for an example of this kind of image, by another male Lenox china painter, Bruno Geyer.

Porcelain Paradise:
The China-Painted Ceramics of Kurt Weiser

by Edward Lebow

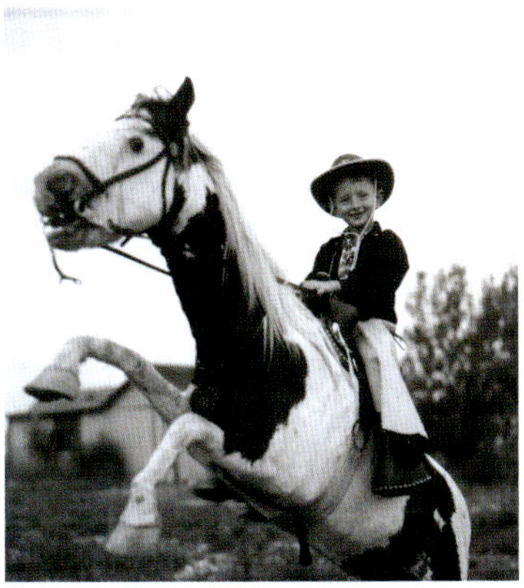

The artist enjoying his childhood in St. Ignace, Michigan, ca. 1956.

James Harner was the kind of inspiring teacher who could conjure a paradise of fantasies for any fourth grader willing to listen. All he needed was his radio-smooth voice, a storybook, and a slice of the school day, recalled Kurt Weiser, a student of Harner's in 1962. After the morning spelling test, Harner would pass out paper and pencils, open a story and have the class draw along with his reading. With scenes from *Smoky the Cowhorse* and other children's book sagas billowing in his head, Weiser would sail away on a voyage of daydreaming and drawing.

It was, Weiser said, somewhat facetiously, the beginning of a very bad school habit. "Any class I was lost in, or wasn't interested in, I just whipped out the paper and was gone."[1] His tendency to fight boredom with daydreams and banality with fantasies easily could have condemned him to the failed life of a child left behind. Instead, it cleared Weiser's artistic path to what he calls his "own little garden." Over the past thirty years, he has made his Shangri-La with some of the most distinctively decorated pottery in modern ceramics. His teapots, lidded jars, vases, cups, and other forms initially explored the limitations and beauty of a craft controlled—and not always kindly—by Mother Nature. Then, in the early 1990s, he chucked his faith in kiln gods and began to china-paint ceramic allegories about his own and human nature. The shift gave Weiser the means to air what had long been rattling around in his head. The resulting works have established him as one of the most inventive potters of his generation and helped to liberate china painting from the pretty pleasures of its hobby-shop exile and the lost vaults of Meissen and Sevres.

An artist who has sought refuge in craft, Weiser has mined the histories of both art and craft. He has extracted influences from a wide variety of Asian ceramics, dipped into the tidy forms of Chinese snuff bottles, and raided the poses, approaches, and stylings of numerous artists and cultural camps. His inspirations range—in no particular order—from Leonardo da Vinci, Heironymous Bosch, John James Audubon, Henri Rousseau, Pablo Picasso, and Salvador Dali to the twentieth-century bug illustrator Walter Linsenmaier, cartoonist Robert Crumb, fashion photography, and more. Yet the ironic bite and visual heft of Weiser's works stem from the surrealist density of the cliché-driven scenarios he wraps around his forms.

Obsessive, refined, and richly storied, his works present implied tales of history, science, conquest, lust, predation, and godly or—worse—spousal retribution follow the thoroughly modern Platonic rule that good people dream what evil people practice. Weiser's images aren't the neurotic sermons that have filled narrative ceramics, but the suggestive narcotic reveries of an errant former altar boy with curiosities to quench and a conscience to ease.

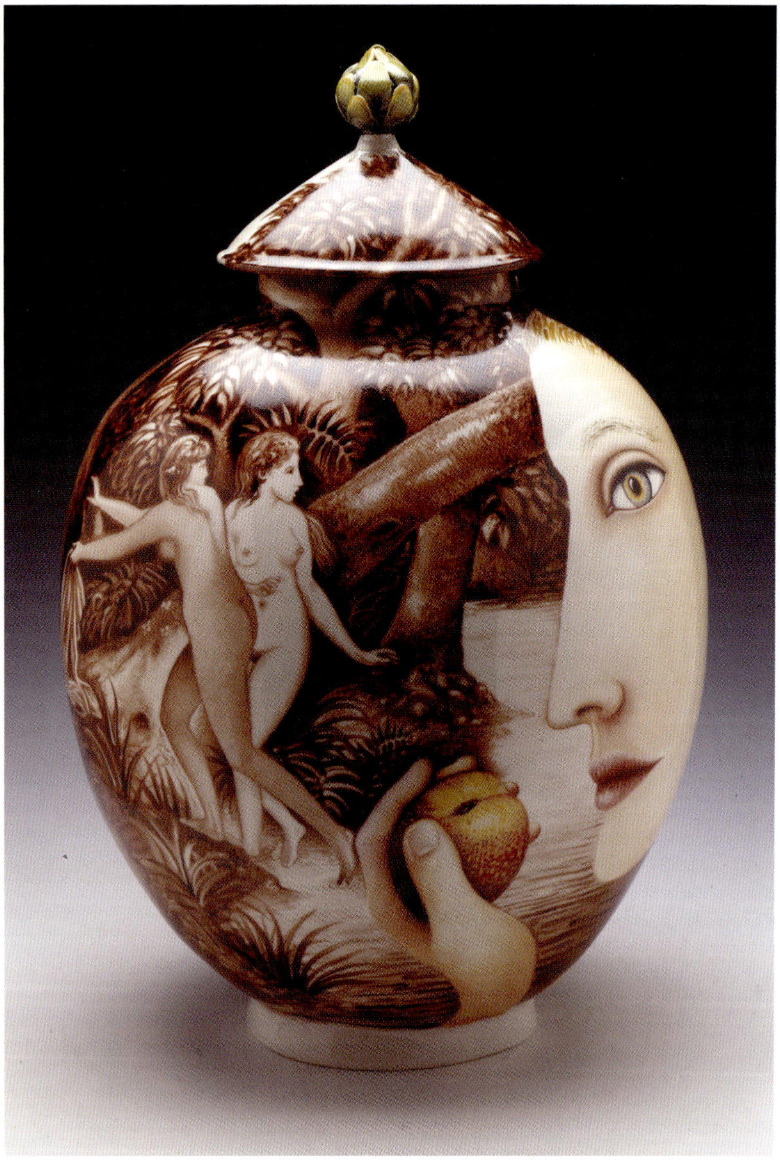

Confidential, 1997
18 x 12 in.
Cast porcelain, china paint

His preferred sedative is the exotic of commonly headlined sins and frailties. It releases him into a private Eden where the temptations of an active libido are always being squelched by sobering reminders of the price of getting caught. Those reminders usually take the form of bugs, birds, and beasts going about their grubby Darwinian business.

There are no clear winners in Weiser's Wild Kingdom, just teeming dualities pushing and pulling in a quiet tug-of-war that belies the visual paradise of his glossy paint job. His counterpoints are fairly consistent: turbulent skies loom over every calm scene; pain is morality's answer to pleasure; rot follows ripeness. And mindless animal urges pave the way to their own special humiliation and misery. Weiser's visual fictions run on an alternating current of real and unreal, and he often provides clues to which is which. Sometimes he distinguishes dreamer from dream by presenting one in living color, the other in sepia, or one as foreground and one behind.

In his lidded jar *Confidential,* 1997, for instance, the one-eyed watcher holding the blushing peach occupies the foreground in full color. An all-knowing sentry of the present, she guards a two-toned background image meant to suggest the past. In this case, Weiser's scene of two naked women at the edge of a wooded stream represents more than a literary past or reference to Susannah and the Elders. It is personal, a fusion of mood and memory that revives a scene he encountered as a kid exploring the woods along the Red Cedar River, several miles from his Michigan home. "I had never seen a woman like that," he recalled. "It was right out of a Rousseau painting." The scene's mixture of fright and titillation opened a window on the secret world of the opposite sex and solidified the link between exploring that off-limits world and the more accessible yet mysterious realm of nature. His efforts to recast such childhood events into fresh adult fictions have nudged his trail of discovery steadily inward, toward some hidden spot where the thrill of experience rings eternally new.

Most of Weiser's ceramic landscapes are set in deciduous swamps and woodlands that resemble wetlands and forests near his childhood home in Michigan, his earliest paradise and all-season studio. Harner encouraged Weiser and his classmates to make the most of their surroundings by paying attention to what they saw, heard, and found outdoors. In one memorable assignment, Harner had the class select, then watch and document a square meter of ground. "I picked a place under an enormous tree," said Weiser. As it turned out, the tree nested an owl, which rained indigestible leftovers—bones, teeth, and fur—from its nocturnal dining on mice and other easy prey in the woods. In addition to these scraps, Weiser noted animal footprints, soil conditions, and new growth; if a storm swept through, he'd record in his diary the kinds of leaves that blew in and settled on his turf.

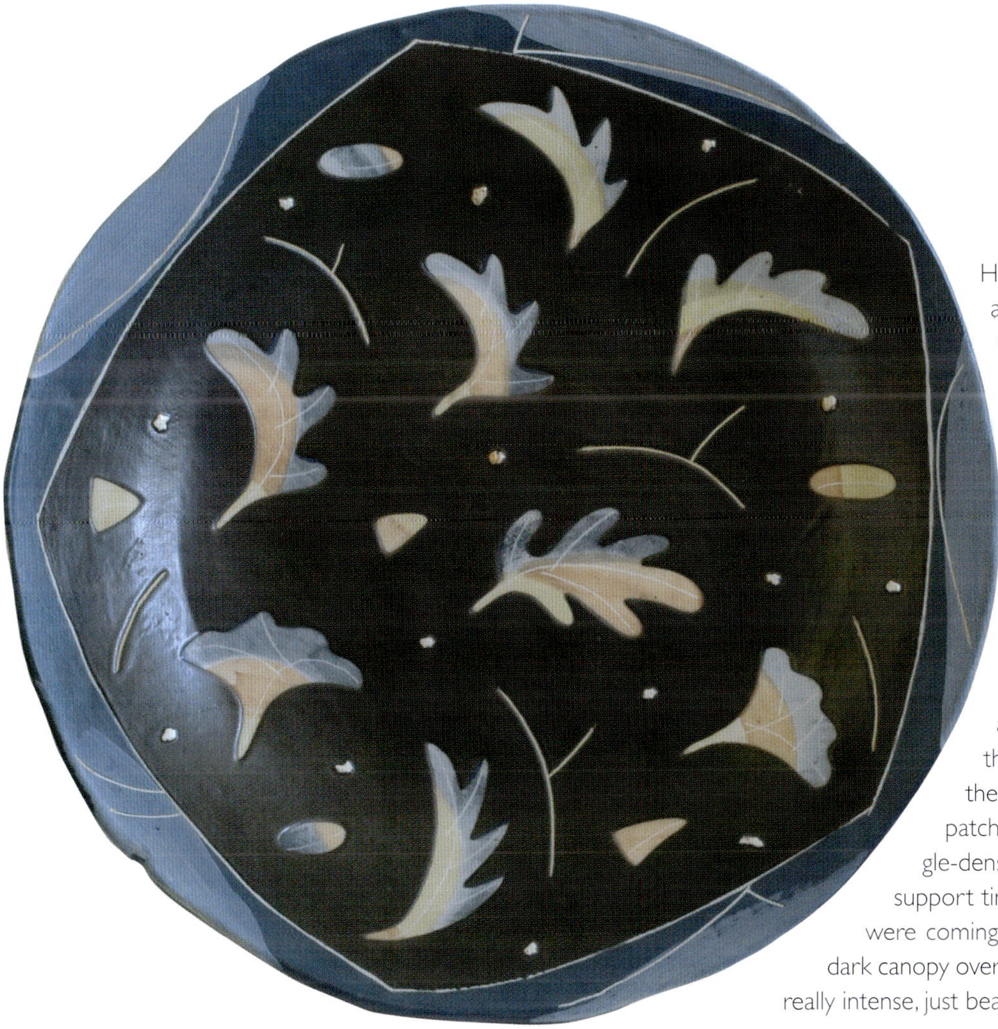

Leaf Plate, 1988
4½ x 27 in. diameter
Glazed stoneware

Harner's encouragement fueled Weiser's already considerable appetite for observation and exploration. Growing up at the suburban edge of the forest, he thrived on the sensual intensity of bushwhacking deep into the woods, building forts, hunting birds and other small game, all the while scanning the underbrush for suitable weapons-grade materials. He described "walking through the woods, your eyes going over every stick, looking for the perfect thing for a bow and arrow, or an animal, or just something that would be a complete surprise." In the latter category he recalls a swampy patch where the thick leafy canopy of jungle-dense trees filtered just enough light to support tiny chartreuse blades of growth. "They were coming out of this black, black dirt with this dark canopy over the top," Weiser remembered. "It was really intense, just beautiful."

Getting a good look at plants and flowers was far easier than examin ing birds and other wild creatures with legs and wings to flee. The only way to get close enough to study wildlife, Weiser discovered, is to "follow them around until they drop dead, or speed up the process a little bit." He discovered Audubon's method of hunting, then studying and painting his prey, in a drawing that showed the great American bird illustrator toting a big gun and a paint can. Weiser recalled that this seemed an ideal way to join his interests in nature and art, "I'd go back in the woods and shoot birds and make these elaborate drawings of them."

The surprise of Weiser's career is that he managed to keep the practice of ceramics separate from the call of wild for as long as he did. Through much of his early career, his creative life was partitioned between pottery and drawing. He filled notebooks with elaborate, stream-of-consciousness sketches that bore no relation to the ceramics he was making. Yet every so often—*Raku Toucan Jar,* 1981 (page 20) and *Leaf Plate,* 1988, for example—he would use clay to wander into the wild. Or, on his large plates, he would apply layers of slips, glazes, and brushwork to create an abstract, illusory depth—of nighttime landscapes or underwater terrain seen through a thatch of free-floating reeds, limbs, or sticks.

These early works are a far cry from the *Field and Stream* surrealism that Weiser began painting on his pots in the early 1990s. Yet they barely show Weiser's inclination to beef up the visual power and layers of his surfaces. They also point to some of the lessons—especially the abstracted naturalism—he had derived from the ceramics of Ferguson, David Shaner, Val Cushing, and, in raku, from Wayne Higby.

Unlike Ferguson and some of his other teachers, Weiser never had to escape the burdensome

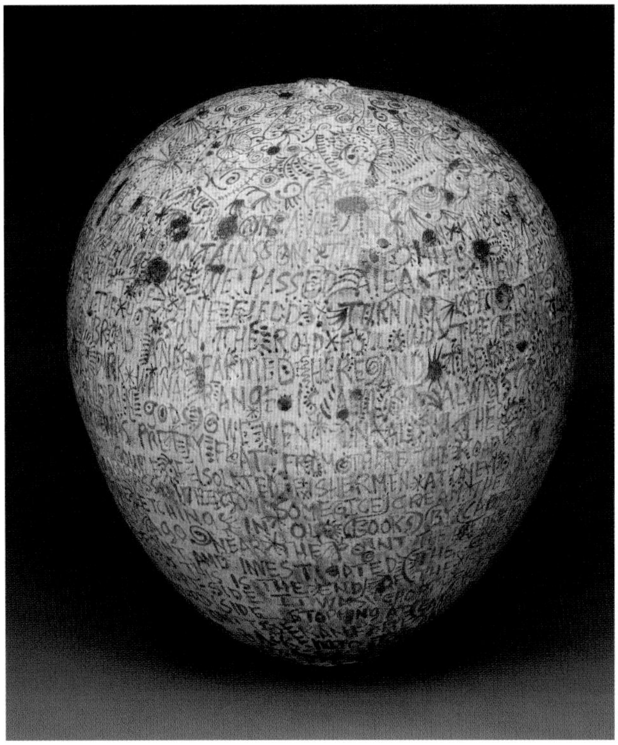

Henry Takemoto (b. 1930), American
Trip to Kaena Point (Hawaii), 1960.
36½ x 33 inches. Coil-built stoneware with underglaze of cobalt lettering through white overglaze. Gift of Lila ColwellCollection of the Jordan Schnitzer Museum of Art

postwar rules of forms prescribed by function that burdened the postwar generation. He always considered his utilitarian ware an opportunity to play with the abstraction of physical proportions. And he never paid much attention to the earnest rhetoric that ornament is a crime. In fact, he assumed that the only purpose of form was to support some sort of fantasy. The trick was to develop forms that could comfortably carry the weight of his wandering daydreams.

Henry Takemoto's decorated pottery from the late 1950s and early 1960s, which Weiser encountered as a student at Kansas City Institute, was the first to convince him that this fusion of form and surface decoration was possible. Beautifully formed and decorated with sprawling, elegant doodles, Takemoto's work epitomized for Weiser the dual realities of pottery's surface and form. When looking at the drawing on Takemoto's pottery, says Weiser, "the form would disappear." If he focused on the form, the drawing would seem to slip from view. Their separate but mutual existence forced the mind to shuttle rapidly from one to the other, alternating between comparing and marveling, moving and pausing, seeing the detail and grasping the whole.

This fantasy of traveling between surface and form played an increasing role as Weiser filled his plates with dense layers of details, and drew from the Bauhaus, natural biology, and art deco to create his mid-1980s vases. Weiser wanted the sensual fire-beaten surfaces of *White Vase*, 1984, *Stepped Vase*, 1985, and *Shino Crackle Cut Foot Vase*, 1984–1985, to look as though they'd resulted from inexplicable natural processes. Yet their sensual appeal extended beyond their stony variegated hues and colorful flashings. *Untitled Vase*, 1984, and *Shino Crackle Cut Foot Vase* have just enough wiggle in their hips and lift in their heels to suggest a spin around the dance floor. Weiser even slipped a heel-like gap under both of these vases to eliminate the heavy visual suction of a fully flat-footed stance. In *Vessel*, 1988, Weiser turned the form into gyrating partners going up in curvaceous plumes of smoke. *Tango*, 1988, is slightly more reserved, supplanting a sinuous dance of clay partners with a constructivist study of parts.

Despite Weiser's efforts to achieve a different kind of sensual ceramic form, the vases, with all of their moving parts, tended to convey the confused sensibility of erotic dancers moving to the beat of a military band. Their oom pah, oom pah left little room for the sensual and narrative interplay that Weiser wanted to incorporate in his work. Part of it, he says, reflected the self-conscious worry of "not wanting to deal with the juicy stuff, or to reveal too much about the things that really interest you."

What really interested him was the exotic backwater of self-absorption that he'd been playing out in his notebooks. To pursue that he not only had to overcome his reluctance to reveal his notebook thinking but also to escape the expectations of conformity that he thought went with directing the Bray. "You were a potter," said Weiser, "...and making pottery in the tradition of these people that had been here before me."

Leaving the Bray for Arizona State University in 1988 was just the escape Weiser needed. He built a studio and a kiln in back of his Tempe house, intending to resume his Tango series. But he didn't get much beyond making a few molds. "I remember doing it and thinking, 'This horse is dead,' " said Weiser. What lived were his fresh impressions of exotic desert flora and

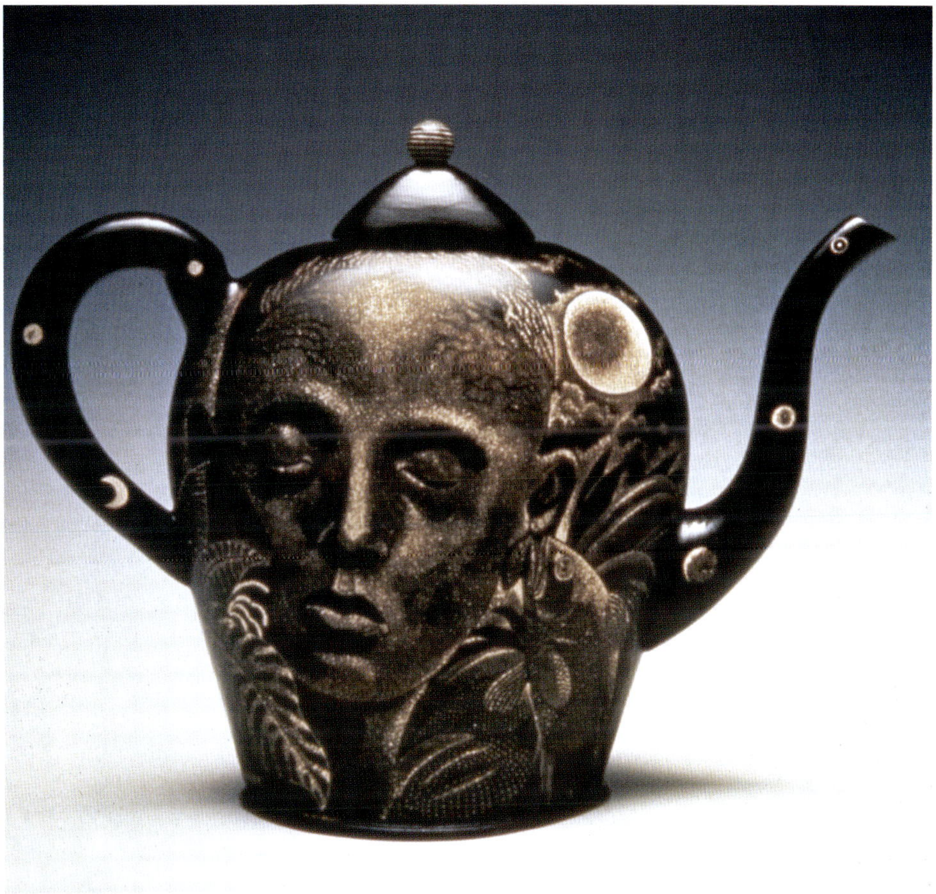

Whispering Fish, 1991
9¾ x 13 x 4½ in.
Cast porcelain, black slip

Untitled Teapot (Scissors in Hand), 1991
11 x 13 x 4 in.
Cast porcelain, black slip

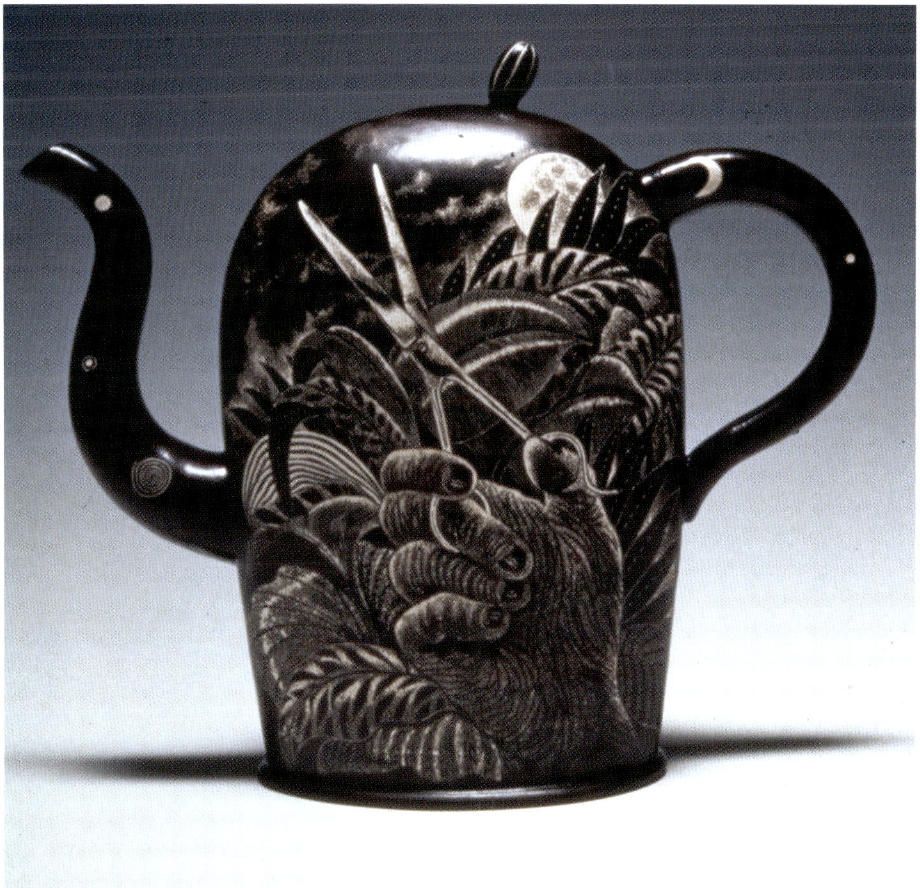

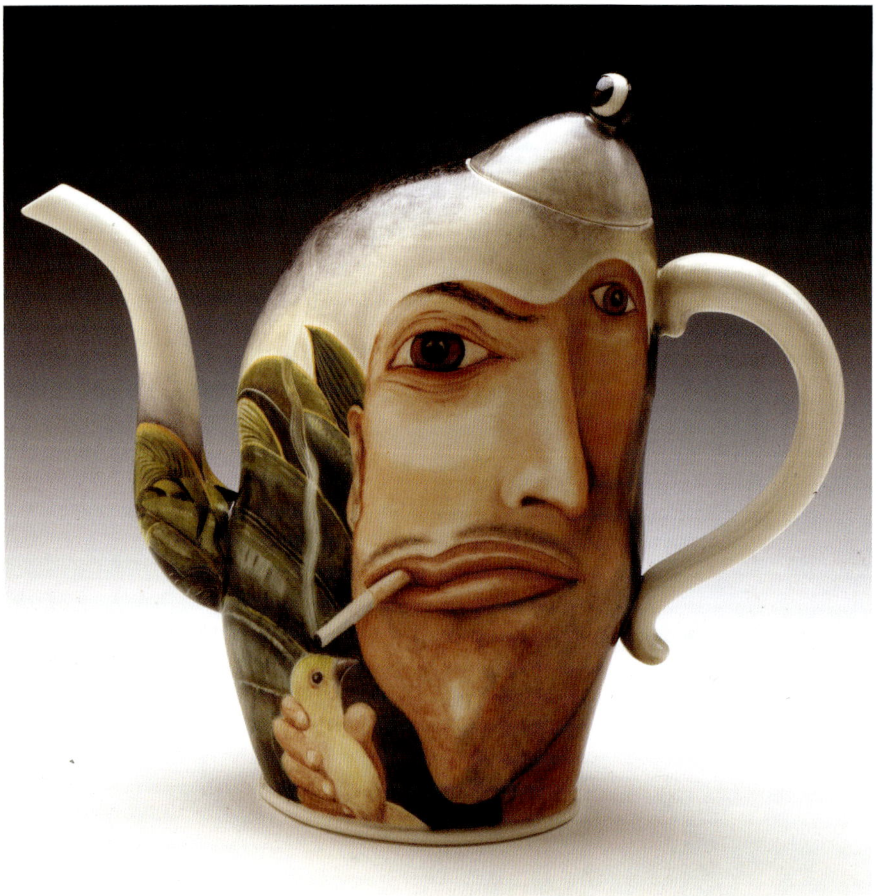

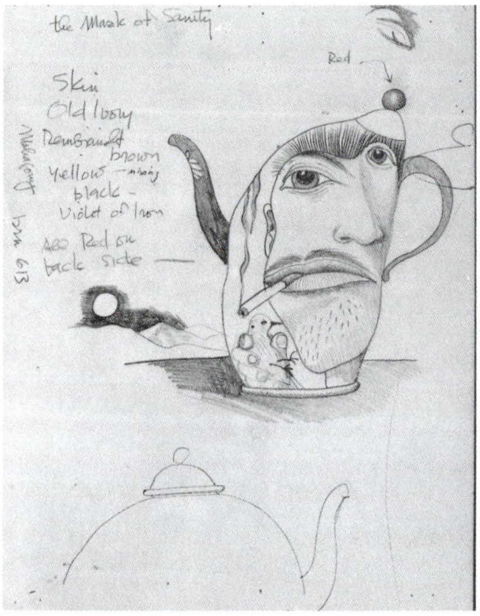

Bird Merchant, 1991
11 x 12½ x 4½ in.
Cast porcelain, china paint

Untitled pencil drawing on paper.
(Study for *The Bird Merchant,* 1991)
14 x 10½ in.

a widening—and frustrating—gap between what he was thinking and what he was making. It was as if the hidden personality of his dream world, a phrase Carl Jung once used to describe himself, was "insisting on coming through to expression."[2] A quick trip to Thailand added more other-worldly images to his fantasizing. However, Weiser remained stumped about how to translate them into clay.

He found the answer at the breakfast table, reading the miniscule flurry of facts, advice, and believe-it-or-nots on the back of a cereal box. "I remember as a kid how absorbing that was," he recalled. "I'd be sucked into it." Weiser's epiphany was to make a teapot as densely packed with visual information as a cereal box. The idea was the equivalent of turning a pot into a television set, said Weiser. From across the room, the T.V. was a sculpture." Turning it on, Weiser observed, made it an experience, a place.

His first attempts were in black-and-white. He created cast porcelain forms like the one in *Whispering Fish,* 1991, covered them with a dense black slip, then scratched botanical scenes into the surface. The directness of the process allowed Weiser both to draw what really interested him and to make work that no longer stood apart from what was going on in his head. He could lose himself in obsessive drawing and mind-drifting—just as he had years earlier in Harner's class. But this time he was leading his own little class.

Initially, due to the heat of summer days in Phoenix, his eye drifted into the desert at night, depicting the landscape's sharp and serrated foliage—agaves, palm fronds, cacti—in an eerie, cool bath of moonlight. He cranked the visual mood from eerie to sinister by adding such images as the shut-eyed face to *Whispering Fish* and the hand-held scissors in *Untitled Teapot (Scissors in Hand),* 1991. The sleeping head—inspired by an image of a marble statue that Weiser found in a book—has the look of a perfect listener, for a fish.

The irony of pairing the exotic with the homey decorum of teapots wasn't lost on Weiser. The

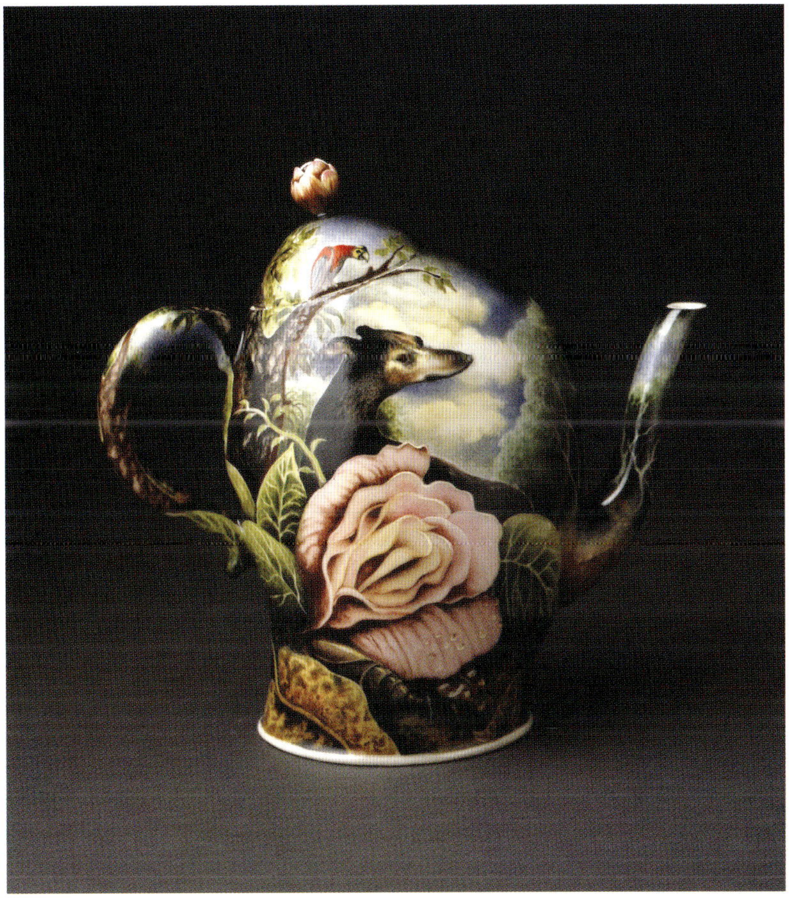

Outdoor Life, 2000
(two views)
12 x 12 in.
Cast porcelain, china paint

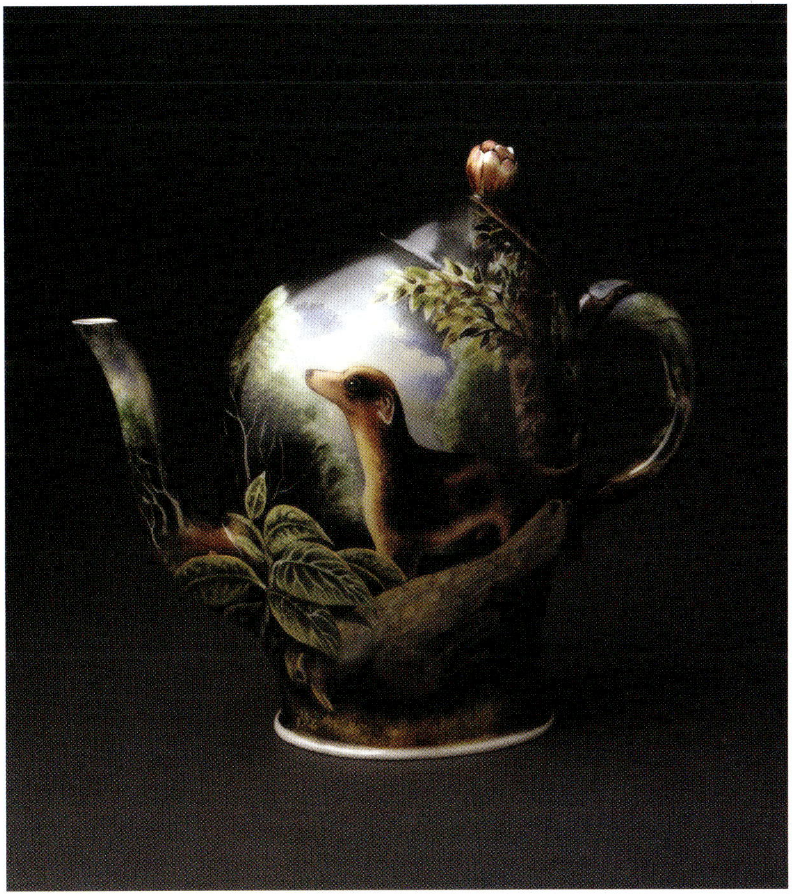

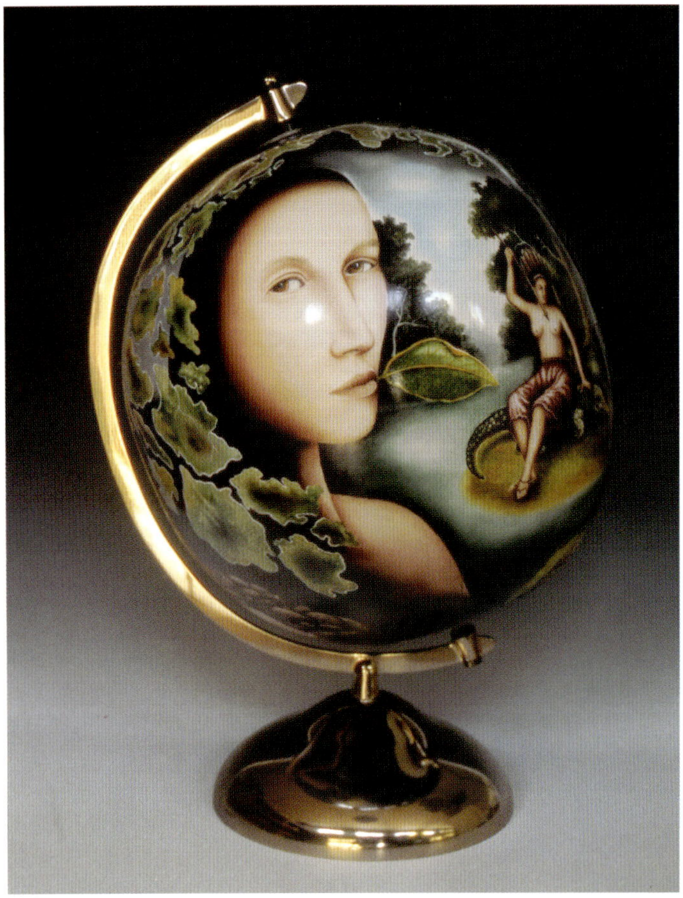

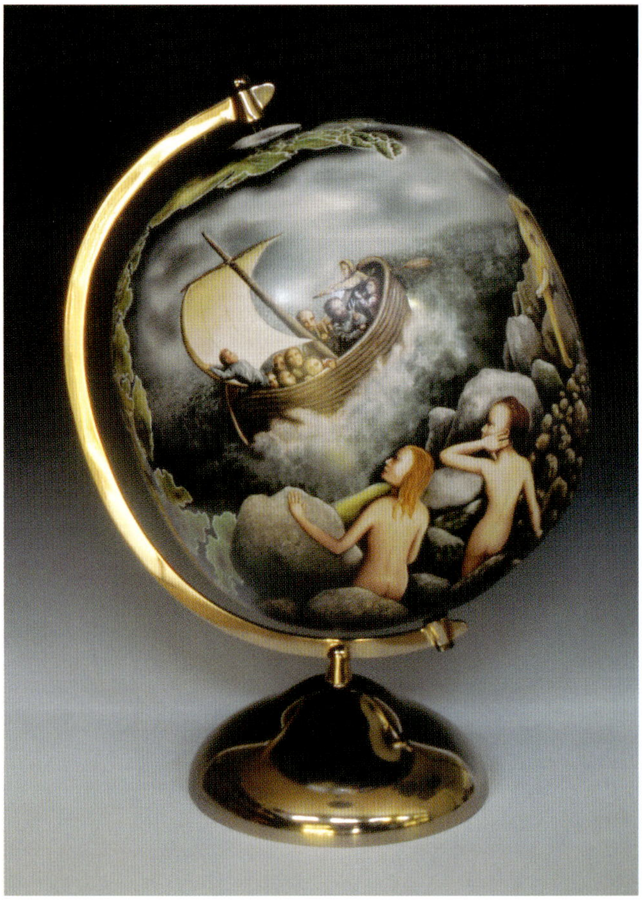

Many Many, 2005
(two views)
25 x 21 in.
Cast porcelain, china paint, bronze

spouted icon of civility gave him the safe cover he needed to unwind his notebooked psyche. Yet turning pots into televisions demanded scenes sufficiently compelling to carry him and viewers away. Weiser began by making the most of the black-and-white scenes, then swapped his sgraffito tool for a paintbrush and some color. As with most new expressions in ceramics, Weiser found the materials by turning to something old: china painting.

Weiser's schooling had all but inoculated him against the decorative stylings of nineteenth-century Meissen and Sevres china. Their tidy little scenic reserves were just the kind of prissy expression that twentieth-century American studio ceramics supposedly had overcome. Yet seeing the direction that Weiser was headed in his black-and-white pots, the late Donald Rabiner, an art historian at ASU, suggested that he take a look at the European masters of china painting.

Weiser was reluctant at first, but he found that china painting was the perfect medium to doodle his daydreams in living color. Unlike glazes, which enter the kiln as pastel-tinted powders and emerge as vitrified colors, china paints start out in full color. For the first time in his career, Weiser could paint on clay without having to imagine what the result would be. He could apply the colors, wipe them off, or build them layer by layer into the hues and visual richness he wanted.

Weiser's handling of colors, figures, and landscapes in *Bird Merchant,* 1991 (p. 34) is fairly rudimentary compared with the lush scenes he would paint in *Iguana,* 1992 (p. 45), *Outdoor Life,* 2000 (p. 35), *Fidelity*, 2000 (p. 54), and other more recent pots. Yet it exemplifies the visual dualities he pursued to give his works their conceptual edge. The oddities are apparent in a glimpse. The merchant's distorted face looms slightly off center like some fun-house drunk. A leafy landscape fills the background and, around the backside, a bird squawks at a hand palming a small round fruit. The vendor's fresh-from-prison face—cigarette dangling from the corner

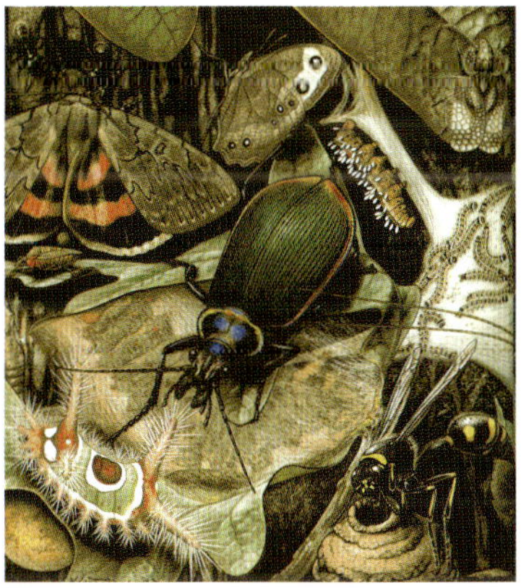

Illustration from Walter Linsenmaier, *Insects of the World.* New York: McGraw-Hill, 1972, p. 14.

of his mouth—seems an unlikely bearer of hope and good cheer, but in another remake of an actual scene—this one Weiser witnessed in Thailand—he offers the little yellow bird in his hand for sale.

In *Bird Merchant* and other forms, Weiser eases the eye into orbit around the scenes by nudging the faces and figures to one side. Distorted as it is, the bird vendor's face takes on the pot's curvature below the handle. In *Iguana*, the arm and wrist of the naked woman soothing the gnarly beast curve up from lower trunk of the form, while her head slopes gently onto the pot's broad shoulder. And in the more recent *Red Leaf,* 1998, and *Abduction Jar,* 1999, and *Many Many,* 2005, Weiser elongated and offset the dominant faces to allow them to fully inhabit their pottery profiles. These efforts to fit the images to their forms not only secured the decorative space he needed to depict the world around his subjects, it also helped to give them the illusion of filling out their underlying volume.

As Weiser became more adept at handling the brush, his forms became larger and slightly more rotund, their scenes richer and thicker with sensual detail. The beautiful overlap of flowers, leaves, forest in *Outdoor Life,* and *Night Harvest,* 2002, reflect the lush layers of the deciduous forests he tromped through as a child. These and other works also reflect Weiser's wide artistic borrowings. His broad-shouldered covered jars are distorted versions of forms derived from combinations of Ch'ing dynasty porcelain containers and Chinese jade snuff bottles. To fill his bill of plants and animals, he lifted details from eighteenth- and nineteenth-century botanical illustrations, bird images by Audubon, and bug pictures by Walter Linsenmaier and others. The marsupial clinging to the branches in *Orbit*, 2001, resembles one that Weiser found in the Museum of Natural History in Paris, the same collection from which Henri Rousseau tapped wild beasts for his jungles.

Weiser drew extensively from the light-filled vistas of the Hudson River School, the turbulent eternity depicted by the German Romantics, and the Good-Friday skies and benumbed-looking visages found in Flemish and Italian Renaissance paintings. The idea for the lady holding the iguana in *Iguana*, for instance, came from Leonardo da Vinci's *A Lady with an Ermine* (1482–83). The cool, contemplative siren gazing out from the *Abduction Jar* is from Rogier van der Weyden's *Portrait of a Young Lady*, 1460. Throughout his works, Weiser used Renaissance-inspired hand gestures as their own sign language to convey the dualities of giving and taking, touching and seizing. He relied repeatedly on popular culture, often Photoshopping faces from fashion magazines to stretch their features across the surfaces of his forms.

Borrowings aside, Weiser's dark, ironic sensibility is entirely homegrown. Conveying plenty about the steamy, chirping, and cawing swamp of his psyche, his story lines remain just vague enough to free him from having to claim responsibility for whatever messages people think they suggest. His narrative gist is Adam and Eve according to Darwin, but delivered in the hallucinatory spirit of Lautréamont's surrealist epic *The Songs of Maldoror*, which Weiser had read in high school. His dual-sided worlds of bugs, birds, and mammals in constant annoyance and conflict seem barely a step removed from Lautréamont's pot-boiled notions of "hares that vanish at the wink of an eye," or "owls bearing rat or frog in their beaks (sweet live food for fledglings)."

Weiser's dog-eat-dog world is awash in odd sexual and scientific tensions between wanting and having, desiring and controlling, seeing and seizing. If the naked figures get the fruit on

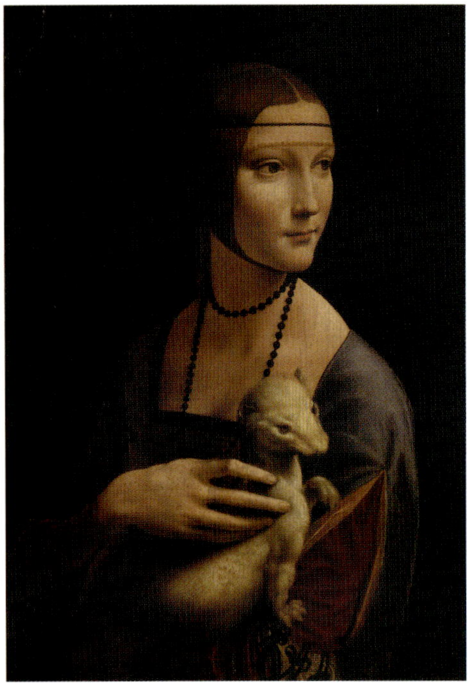

Leonardo da Vinci
Lady with an Ermine, 1482–83
Princess Czartoryski Foundation at the National Museum in Krakow, Poland.

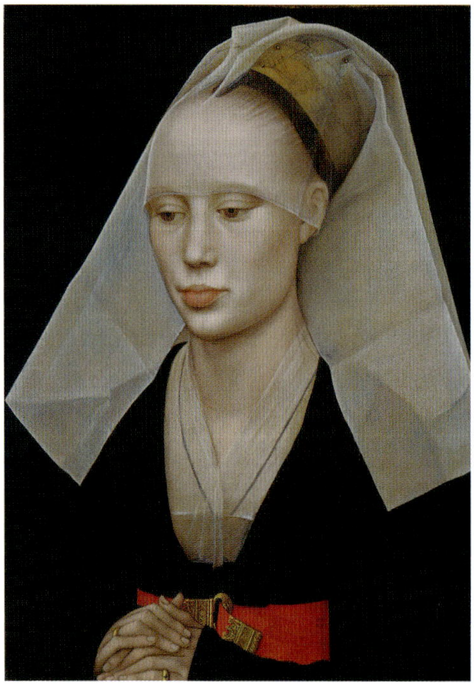

Roger van der Weyden
Portrait of a Lady, c. 1460
Oil on panel, painted surface: 34 x 25.5 cm, panel: 37 x 27 cm, Andrew W. Mellon Collection, National Gallery of Art, Washington D.C.1937.1.44

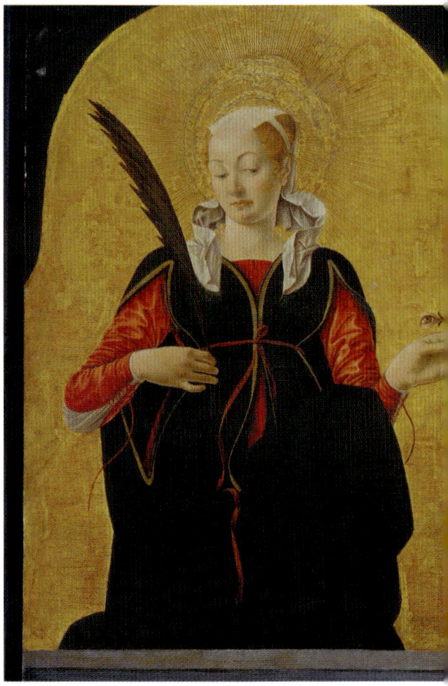

Francesco del Cossa
Saint Lucy, c. 1473/1474,
Tempera on panel, 77.2 x 56 cm., Samuel H. Kress Collection, National Gallery of Art, 1939.1.228

one side of his pots, someone or something is usually paying for it on the other. In *Iguana*, for instance, the naked lady stroking a reptile is offset around the back by a snake in the jungle snatching a pitiful-looking lizard. In other works, he simply plays good dog against bad. In his sensational teapot *Outdoor Life*, a hungry-looking doberman cranes his neck to catch a whiff of the genital-like petals of a voluptuous pink flower, while on the other side, a dutiful beagle, the American icon of obedience, patiently awaits its next command.

The beauty of these works is in the way Weiser's compositional lines seem to arise and move seamlessly from the topography of his forms. In *Outdoor Life*, as in most of his other teapots, those lines sweep across the body of the vessel, carrying the eye from handle to spout. In the case of his heavily foliated covered jars, such as *Night Harvest*, 2002,, the images move easily around the curving walls, one side dissolving with oceanic ease into the next. In the two-piece *Semi-Conscious,* 2002 (p. 65) and other twin-vessel pieces, Weiser cues the eye to leap from lobe to lobe.

The increased complexity of these forms recall the twin-vesseled tango pots he made during his last years at the Bray. But this time, he leads the eye through a visual dance that covers the space of his forms in ways that the reserve painters of Sevres and Meissen probably never imagined. Weiser's directional signs are fairly simple. In *Night Harvest,* for example, a tilted head moves the eye along. In *Bright Angel*, 2001, the draped legs of a background figure (emerging from just below the ear of a dreamy-looking face) give the nudge. And in *Outdoor Life*, the craned neck of the leering dog and the serpentine tree limbs around him play the visual traffic cops. By compelling the eye to orbit his forms, Weiser conveys a powerful visual equivalence between seeing and moving, mobility and dreams.

His mastery of this virtual world has enabled him in recent years to expand his painted dreams from pottery forms to distorted globes with scenes that revolve—wobbles and all—fully in the

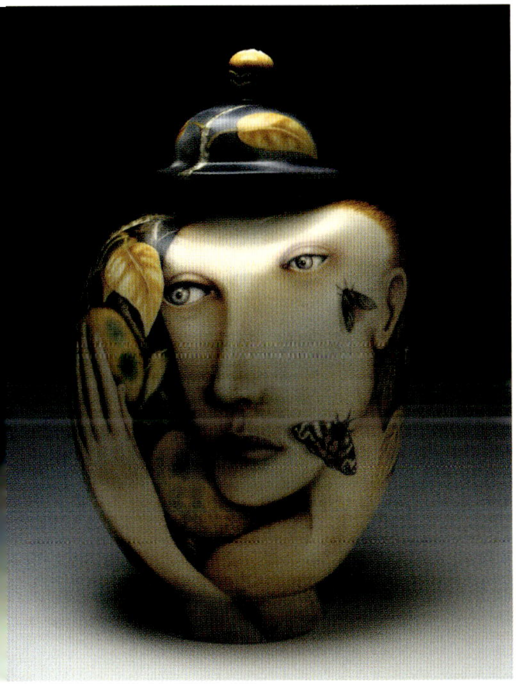

Night Harvest, 2002
17¾ x 12½ x 9 in.
Cast porcelain, china paint

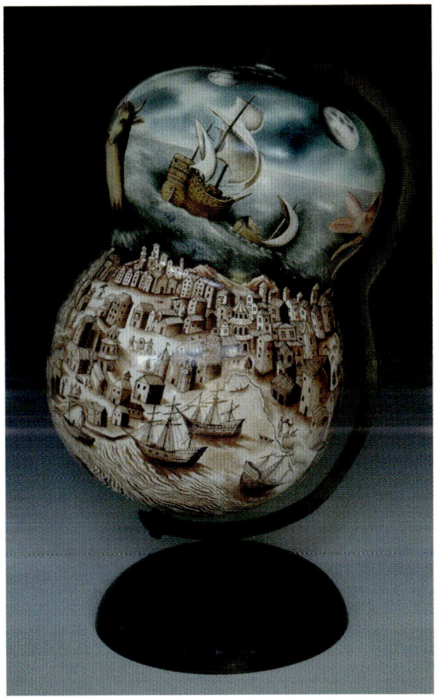

Europa, 2005
27 x 17 in.
Cast porcelain, china paint, bronze

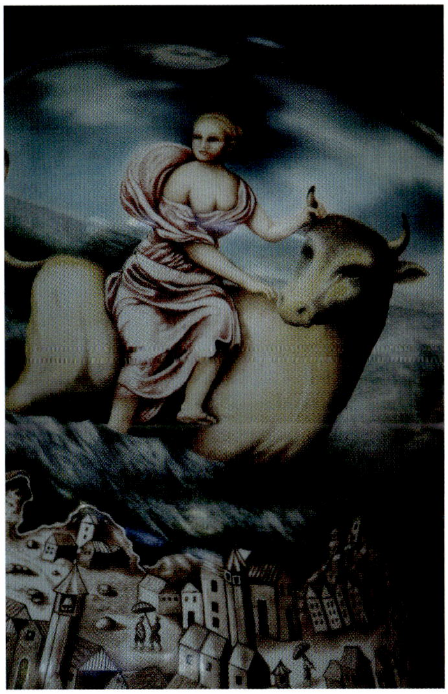

Europa (detail)

round. Gone are the handles, spouts, necks, lids, and other pottery parts that helped him change and subdivide the visual flow of his scenes. Despite his adoption of continuous round forms, Weiser's new world order remains as polarized as ever. The globe of *Europa*, 2005 — cinched at the waist like a Mr. Peanut and divided into water and land, myth and history. Its full-masted ship at sea and land-lubbing hacienda boys—outfitted with wide-brimmed hats and sticks to thrash their victims—hint at the age-old conflict between colonial conquest and native control. However, as often happens in Weiser's work, this whiff of the political quickly leads to the physical. With the firm grip of one hand, Europa has taken the bull by the horn.

He repeats this theme of tamer and the tamed in *Many Many*, 2005. One side of the ovoid globe features a bare-breasted, spear-wielding heroine having her way with a reptile. The other shows an Escher-like ship of fools booming along in a gale through rough seas. With the poor naked innocents waving from the rocks, it's easy to concoct a tale about civilization's arrival—here comes trouble—to despoil the New World. Yet, given Weiser's penchant for daydreams, this might be merely another of Weiser's many voyages to the shores of his exotic nature. As with all of the oceans he paints, this one points to his own little inland sea, where tides of fantasies always seem to be rising into view.

NOTES

1. Unless otherwise noted, all statements by the artist are from interviews with the author, June and August 2006.

2. Carl Jung, *Memory Dreams Reflections,* trans. Richard and Clara Winston. New York: Vintage Books, 1989, 185.

Artist's Statement

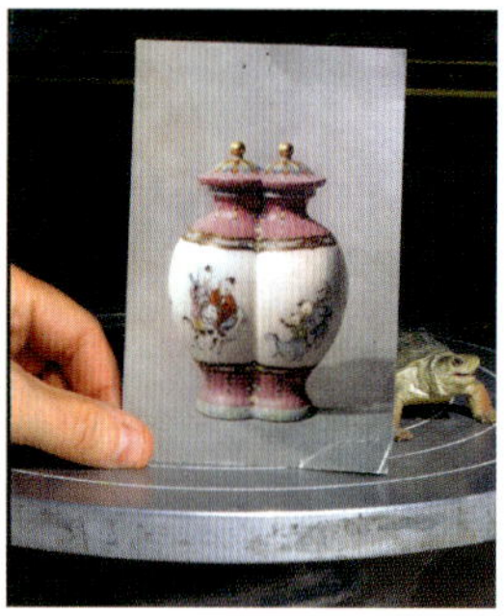

FOR YEARS THE WORK I DID IN CERAMICS was an effort to somehow express the formal and beautiful nature of the material. As interesting as this exploration was, I always had the vague feeling that the best expression of the material only came as a gift of nature: The problem was, nature and I never got along that well.

Somewhere in the midst of this struggle I realized that the materials are there to allow you to say what you need to say, not to tell you what to say. So I gave up trying to control nature and decided to use what I had learned about the materials to express some ideas about nature itself and my place in it.

ABOVE: Postcard sent to the artist from Akio Takamori of historical Chinese double vase. This image inspired Weiser in his own work. BELOW: Kurt Weiser in his studio, 2007

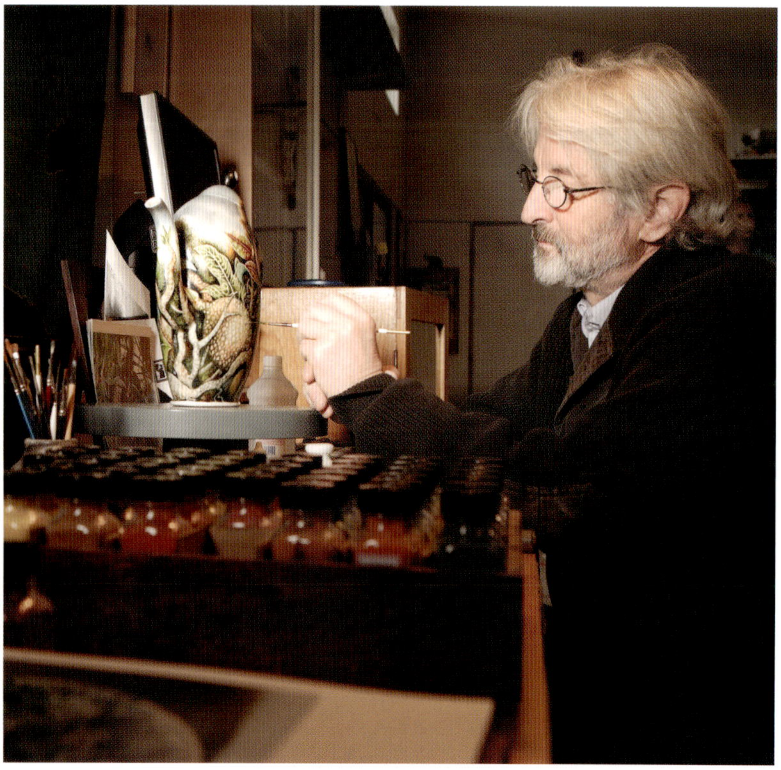

I started drawing and painting on pots out of desperation more than anything else, in order to have some say over what happened on the surface.

I thought about the childhood experience of sitting at the breakfast table staring into the images on the cereal box. Still half asleep, this cereal box absorbed my attention like a sponge. Every bit of information and image held my attention to the exclusion of everything else. I wanted to make something that had the same absorbing and compelling surface that that breakfast cereal box had when I was a child.

At first I made many tea pots and drew on them or painted with china paints. Later on I moved on to other shapes that were mostly classically Chinese but as if viewed in one of those fun house mirrors at a circus.

The ideas and subjects of these paintings on the pots are for the most part just a collection of my own history of fantasy and view of reality. They are built the same way we dream: Around a central idea, a cast of other characters and environments just seem to show up to complete the picture.

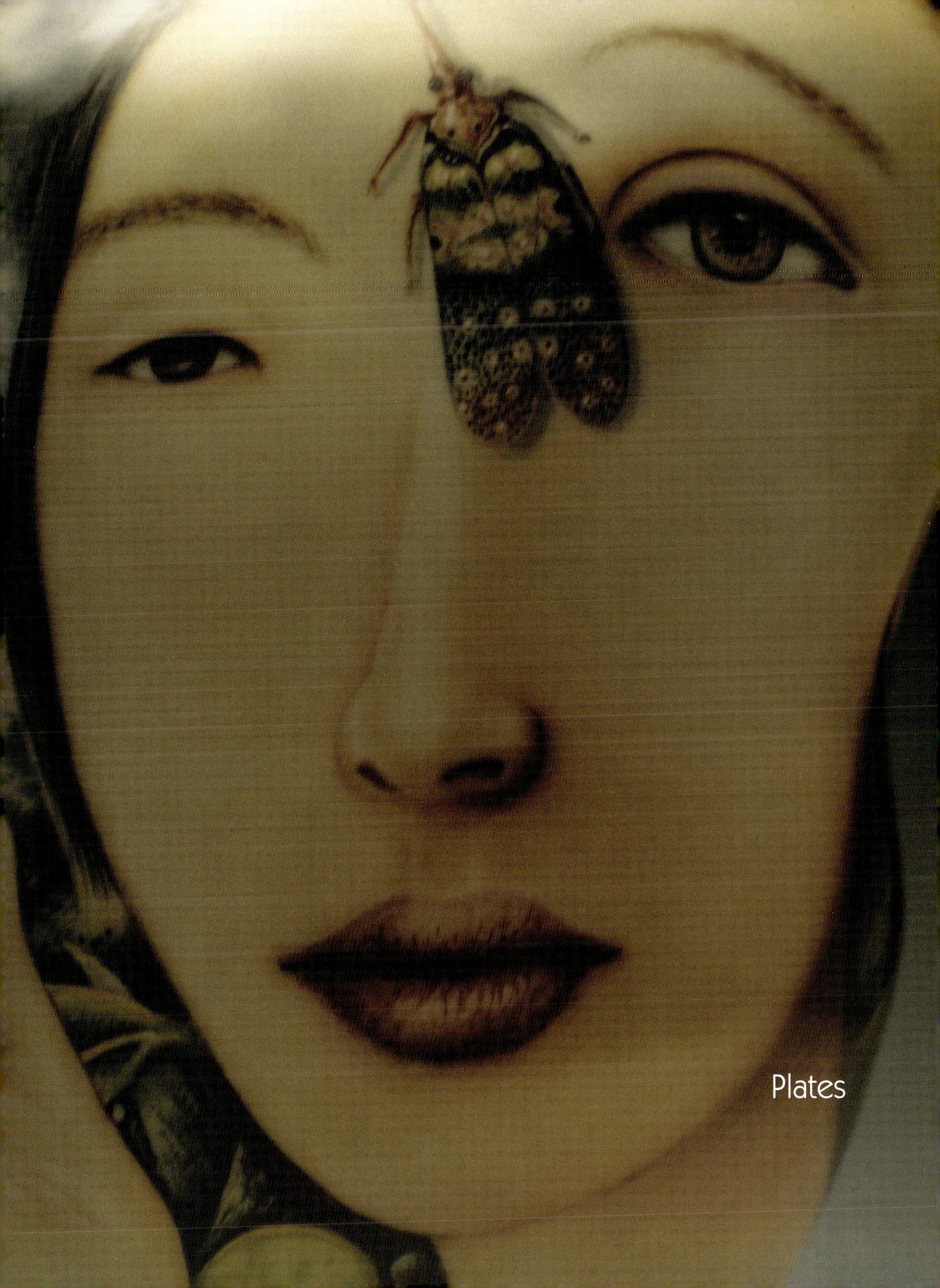
Plates

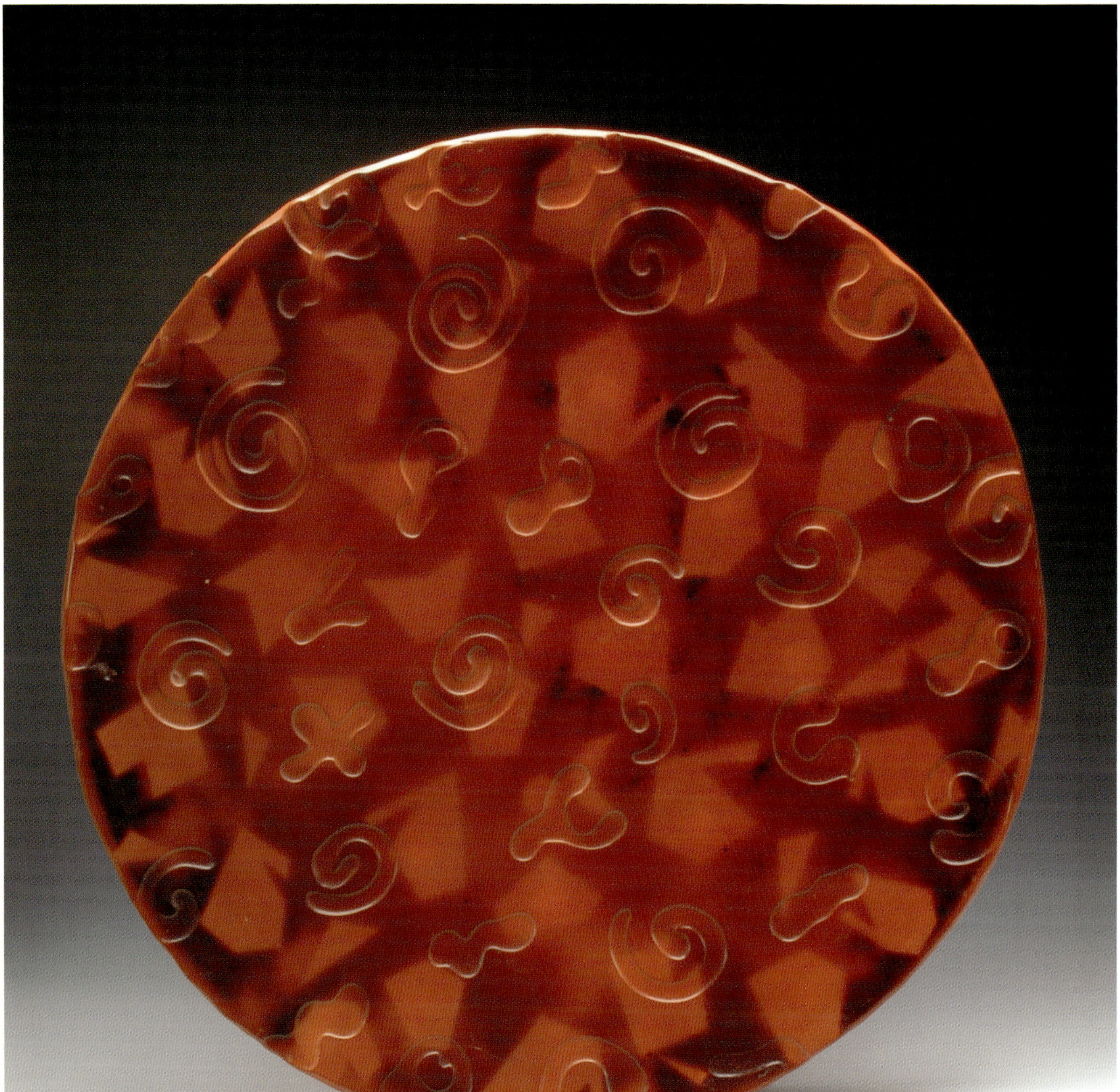

1

Plate, 1982

26½ in. diameter

Earthenware, residual salt-fired

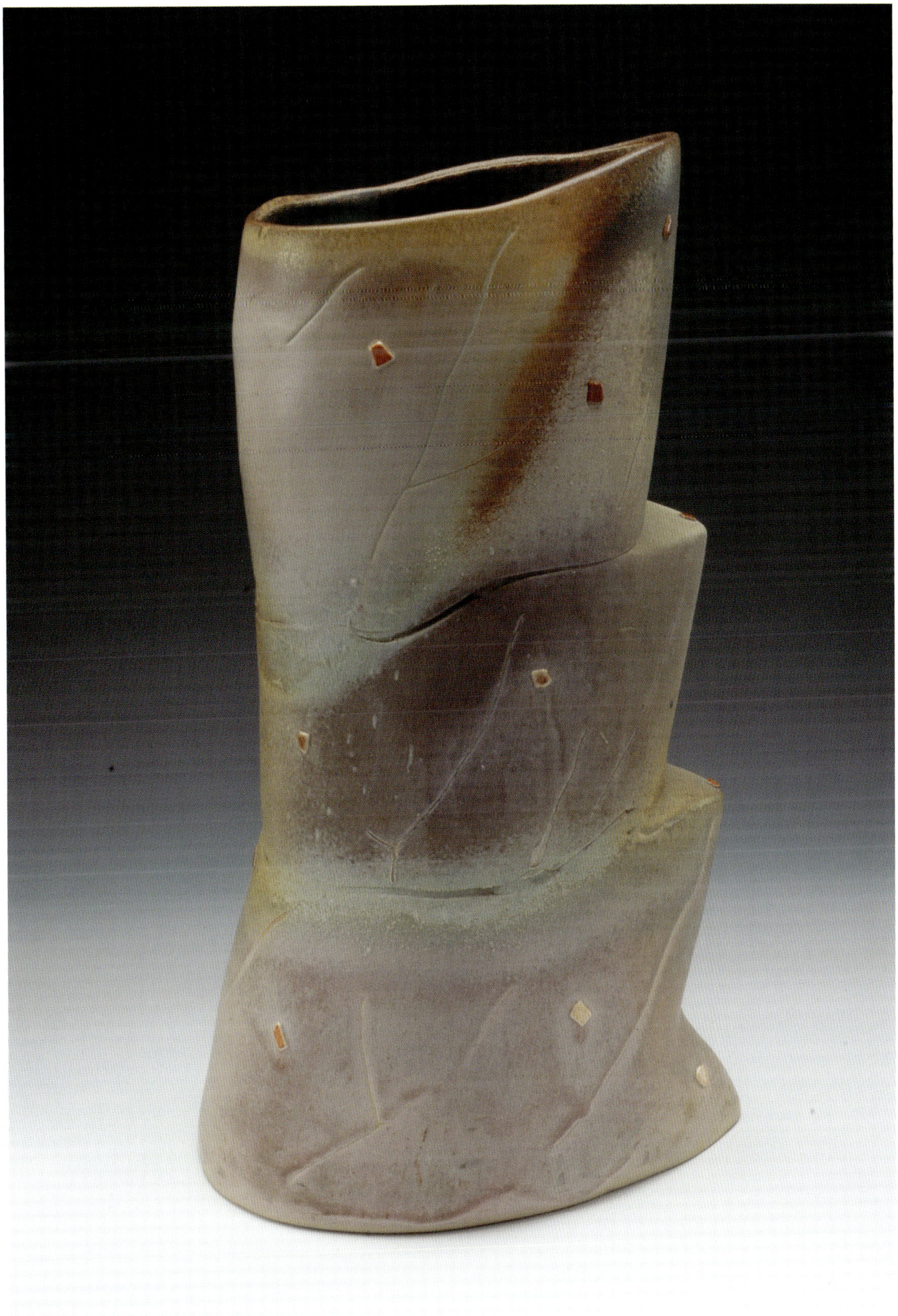

2

Vase, 1984

17½ x 10 x 7¼ in.

Glazed stoneware

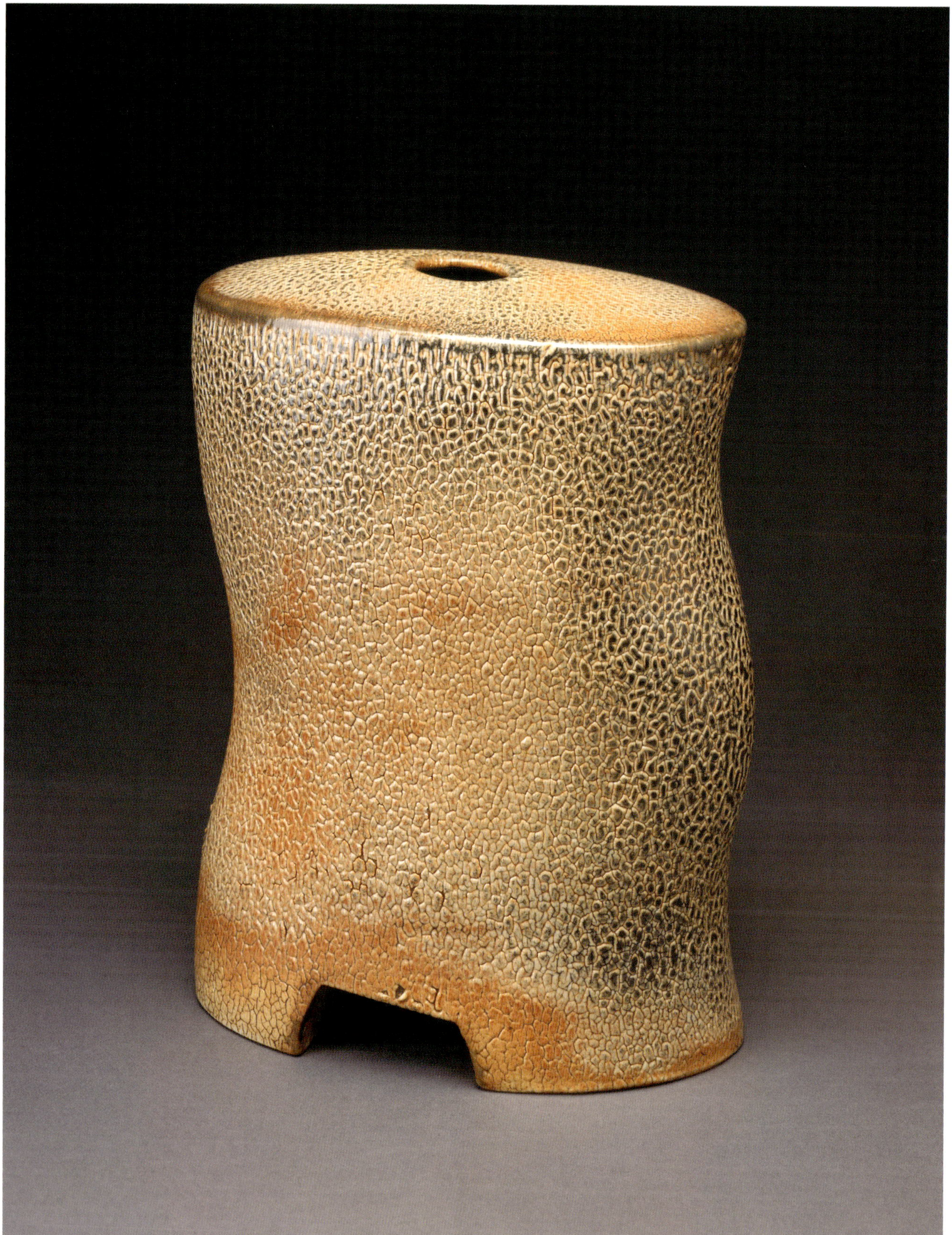

3

Shino Crackle Cut Foot Vase, 1986

12 x 8 x 4 in.

Glazed stoneware

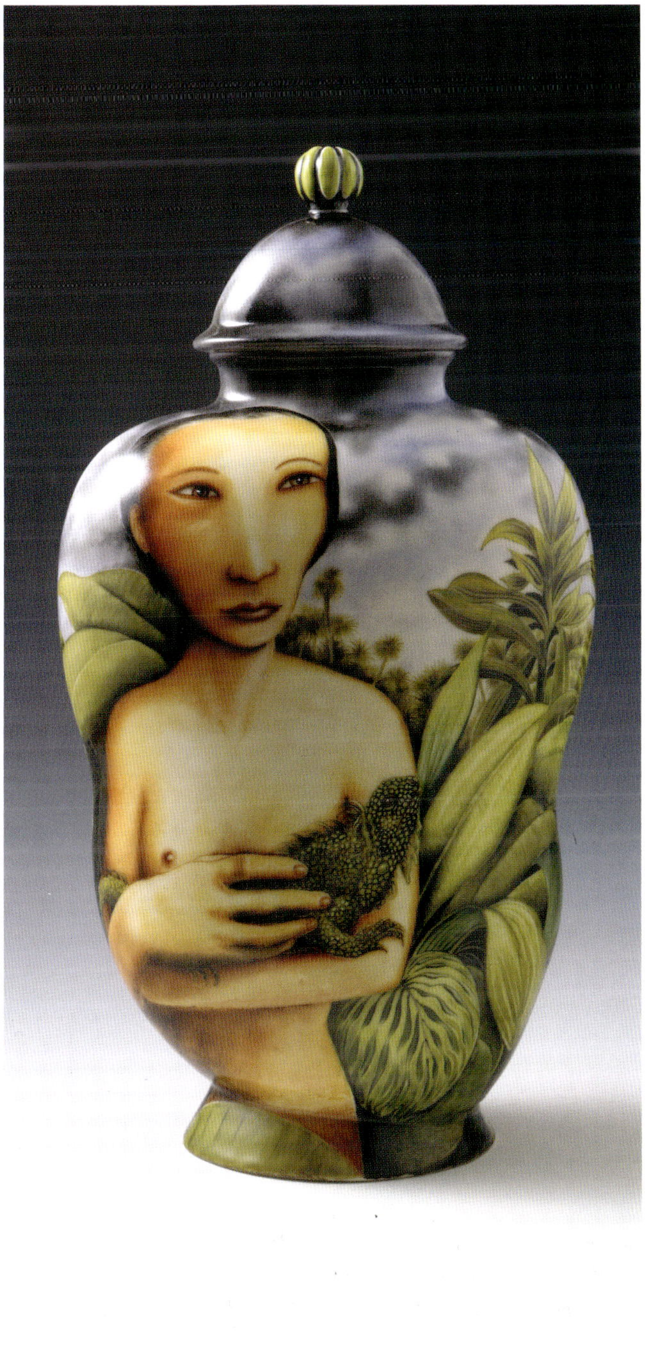

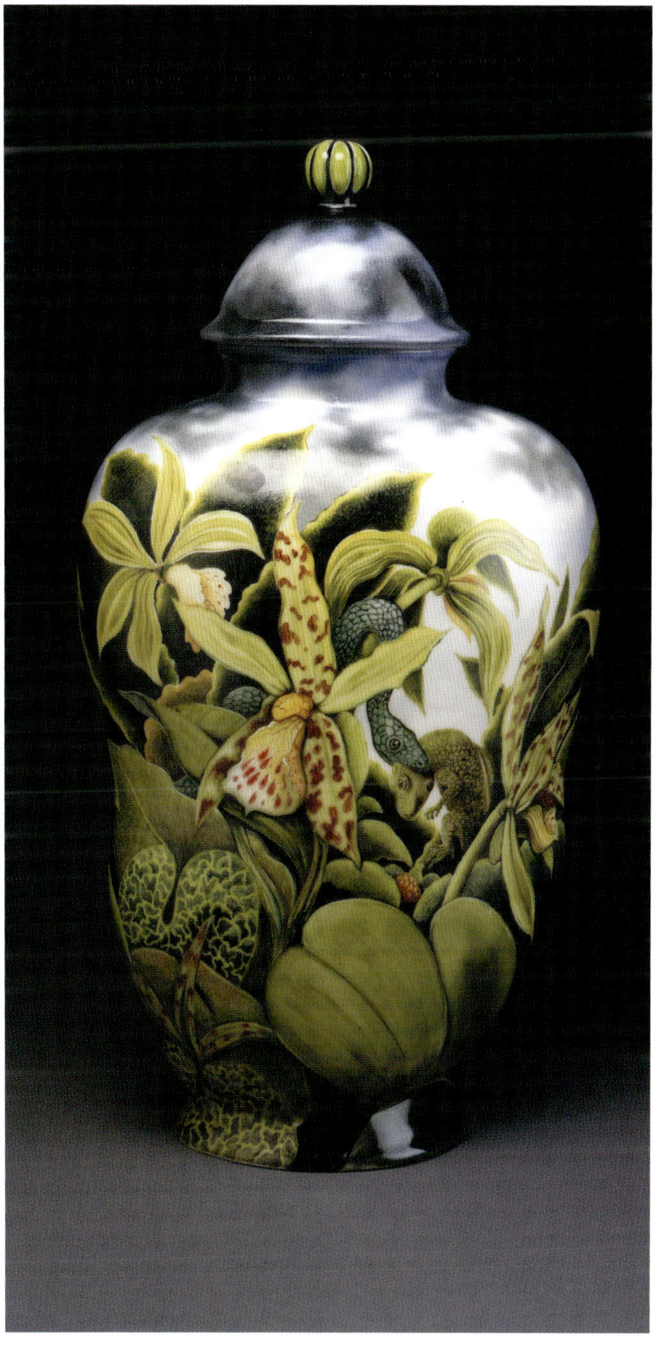

4

Iguana, 1992 (two views)

17¼ x 9 x 5 in.

Cast porcelain, china paint

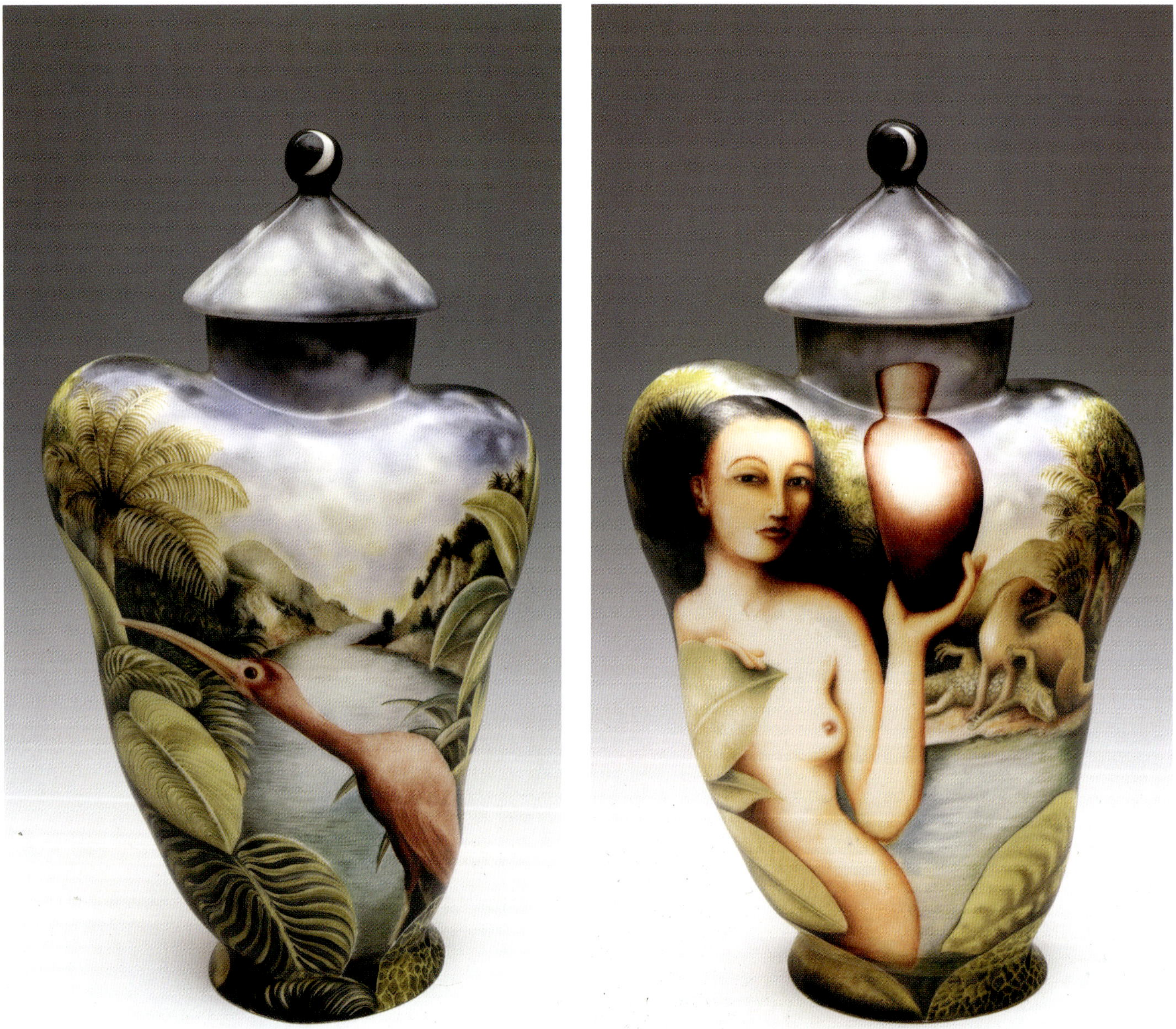

5

Lidded Jar, 1992 (two views)

17 x 10 x 6 in.

Cast porcelain, china paint

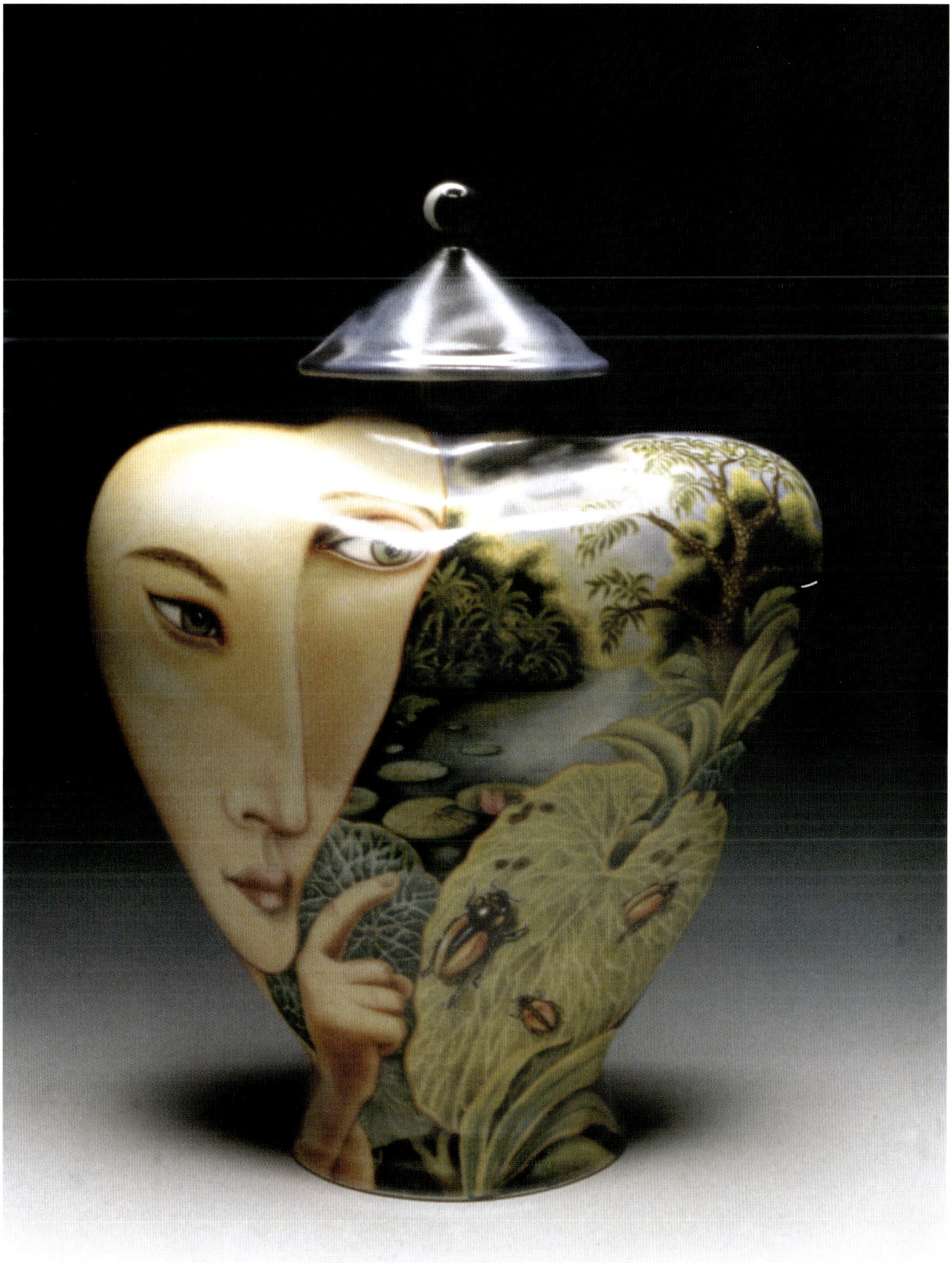

6

Bug Lesson, 1994

17 x 12 in.

Cast porcelain, china paint

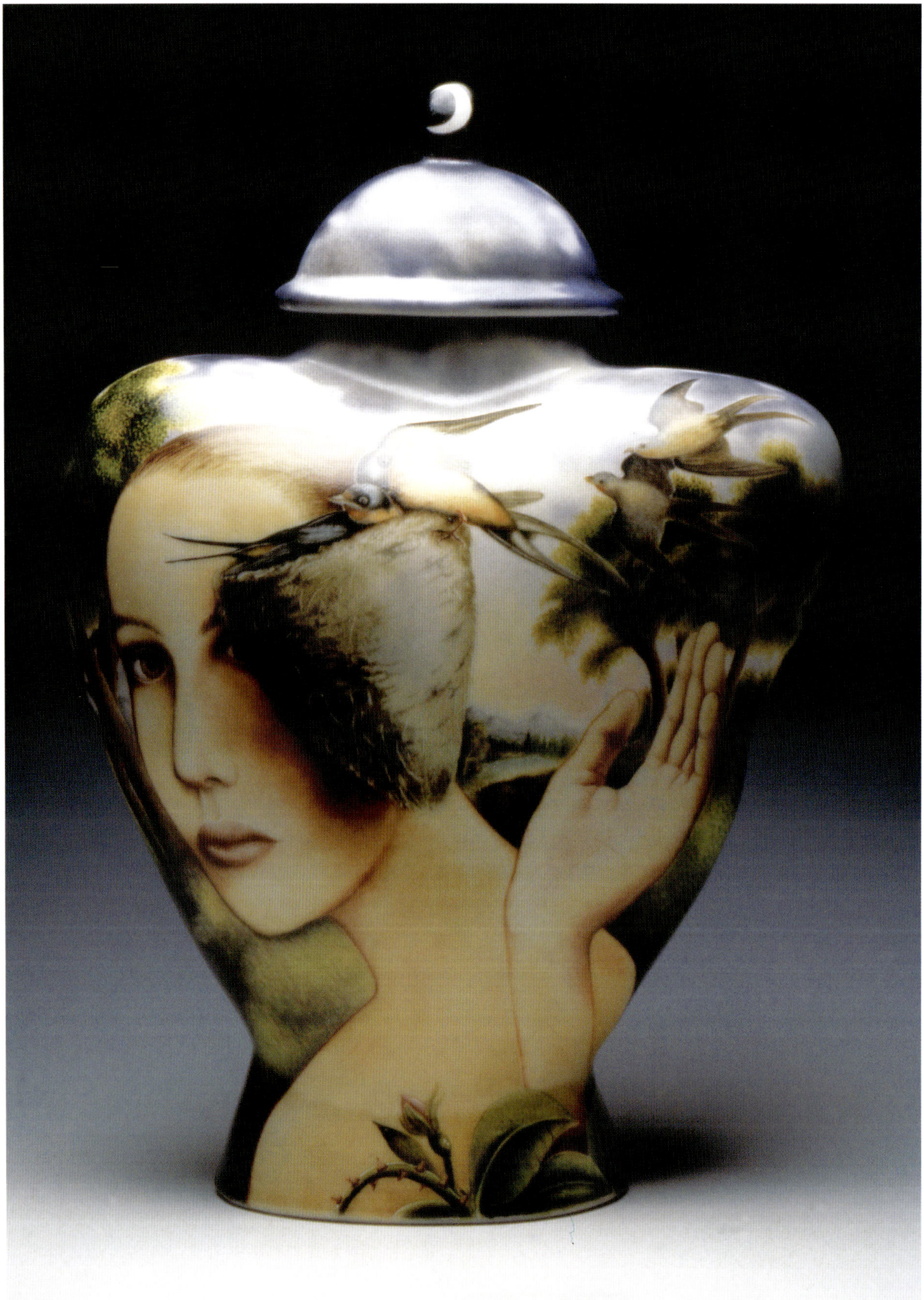

7

Sleeper, 1994 (front)

20 x 12 x 6½ in.

Cast porcelain, china paint

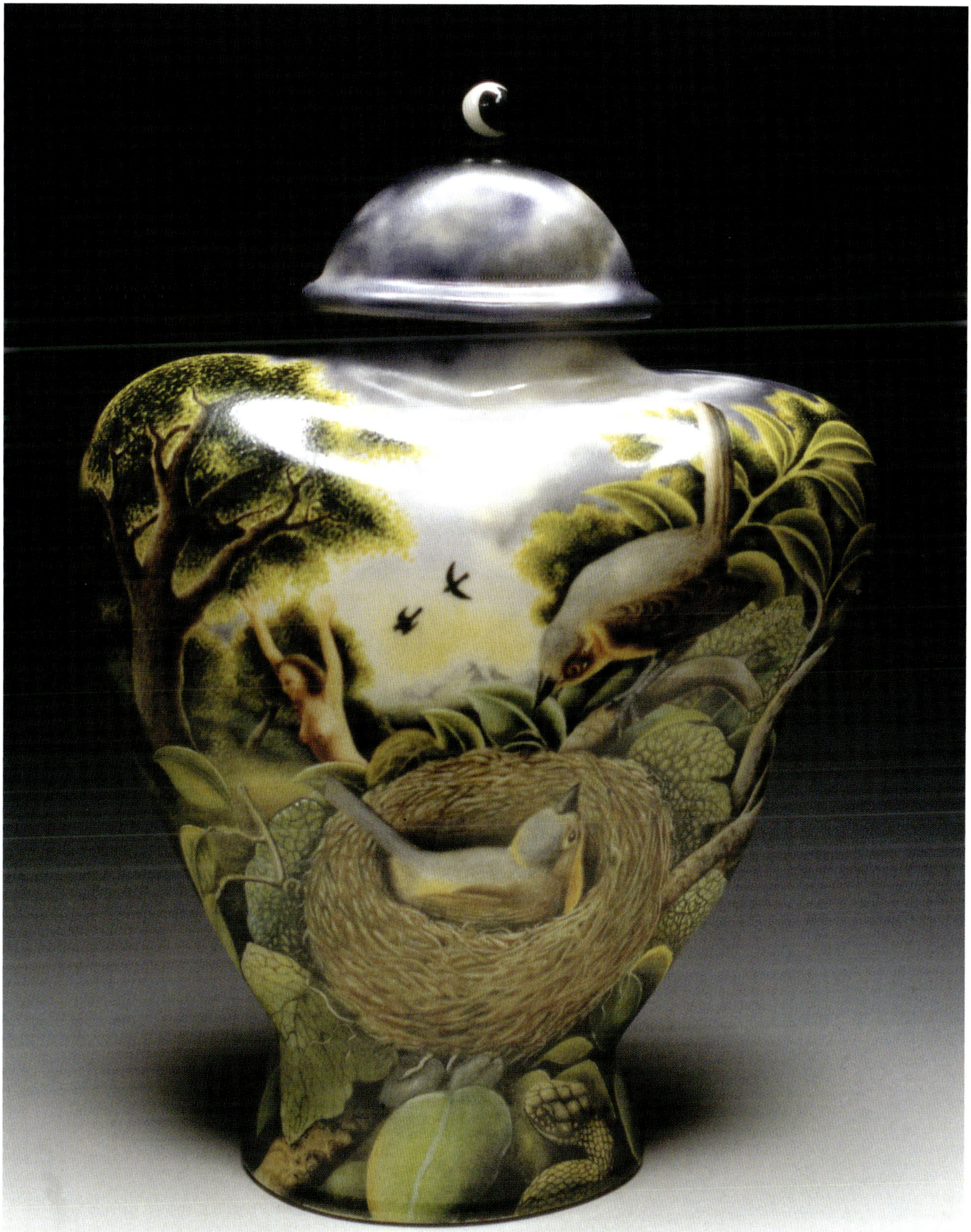

7

Sleeper, 1994 (back)

20 × 12 × 6½ in.

Cast porcelain, china paint

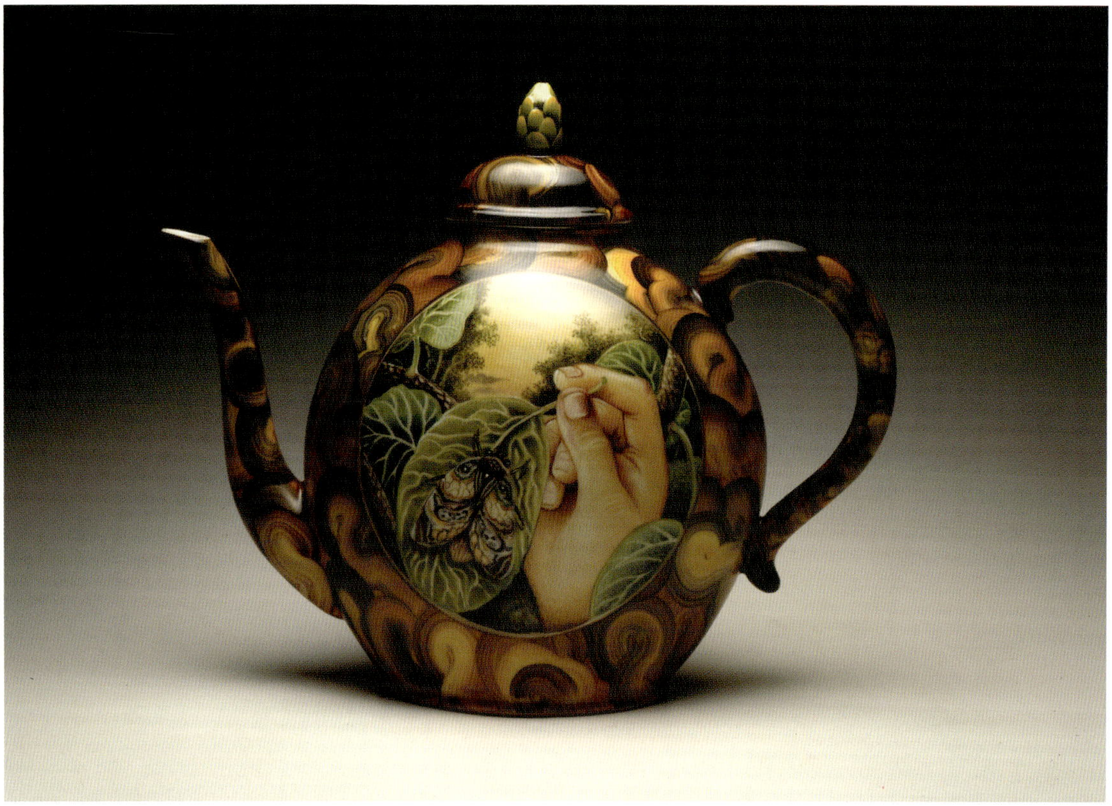

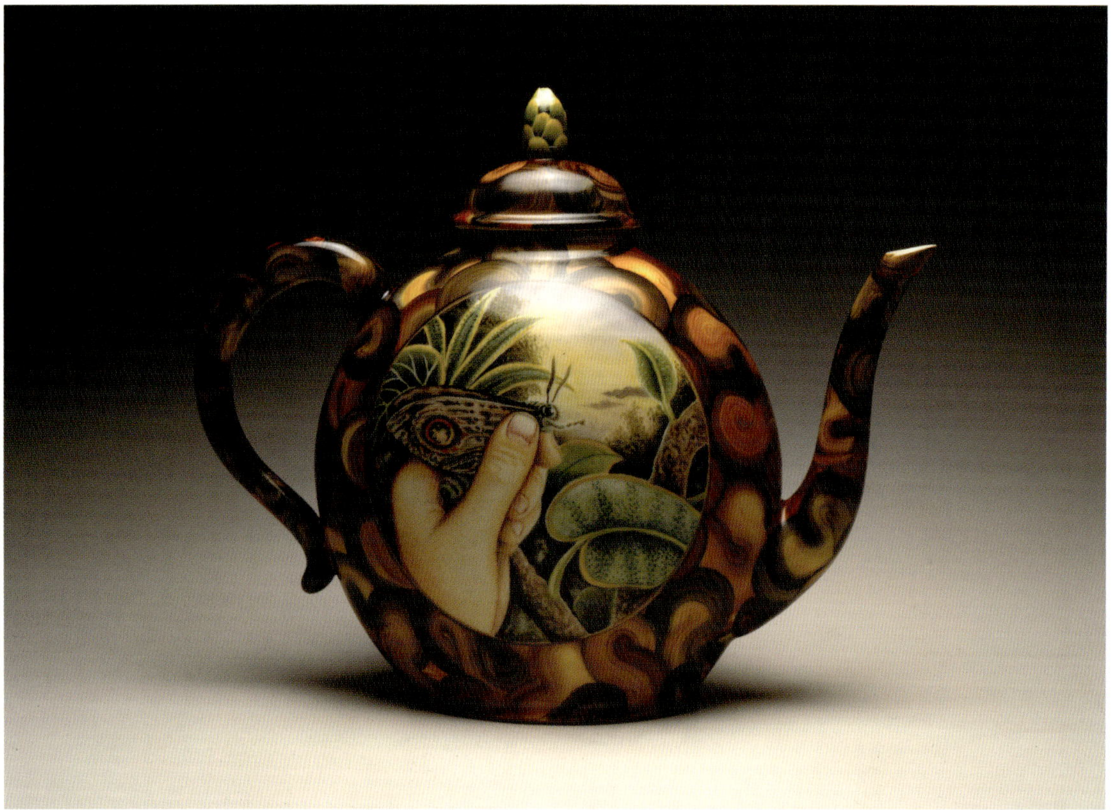

8

Marbleized Teapot with Hand, 1996 (two views)

11¼ x 14¼ x 5 in.

Cast porcelain, china paint

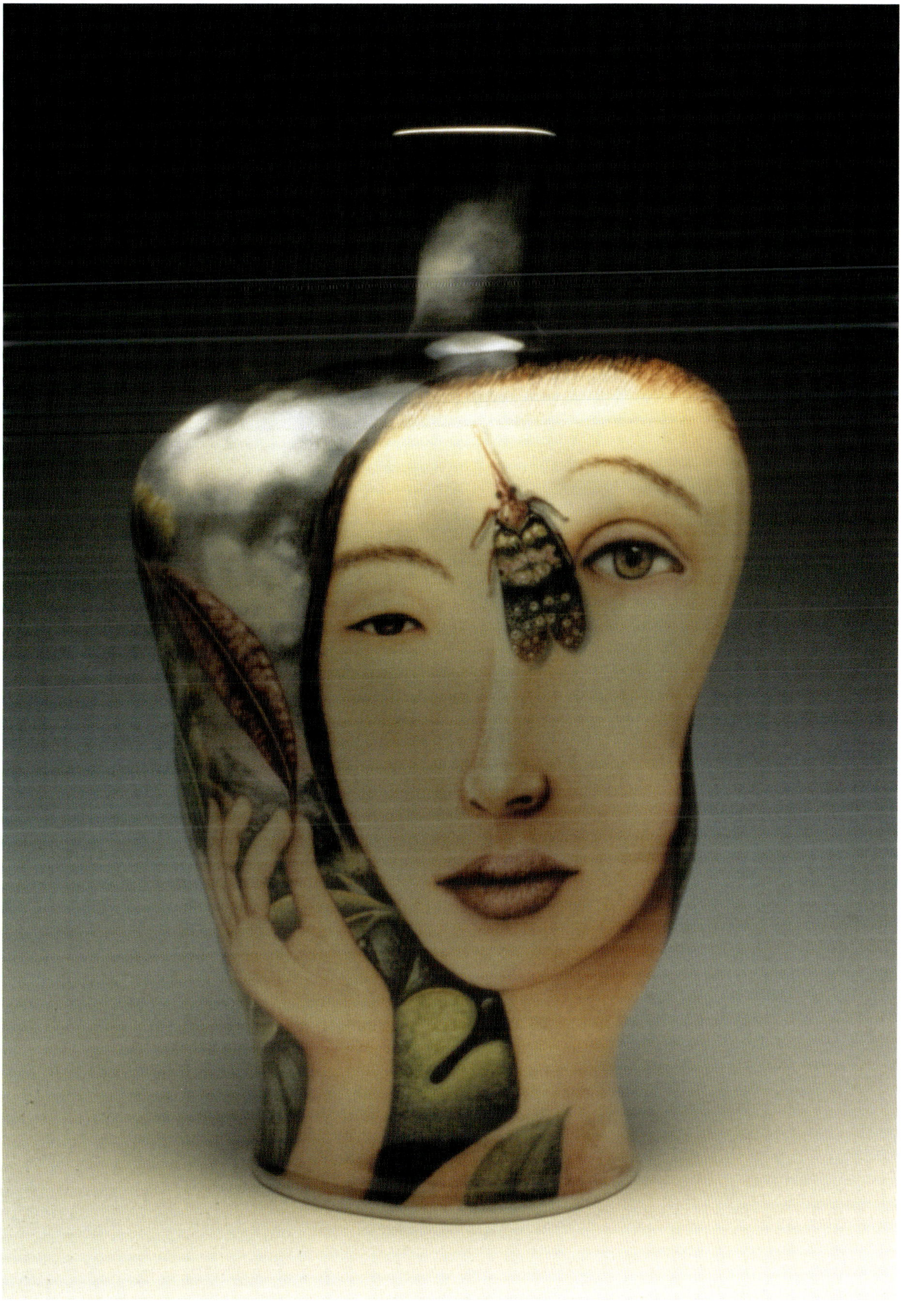

9

Red Leaf, 1998

18½ x 10 x 7 in.

Cast porcelain, china paint

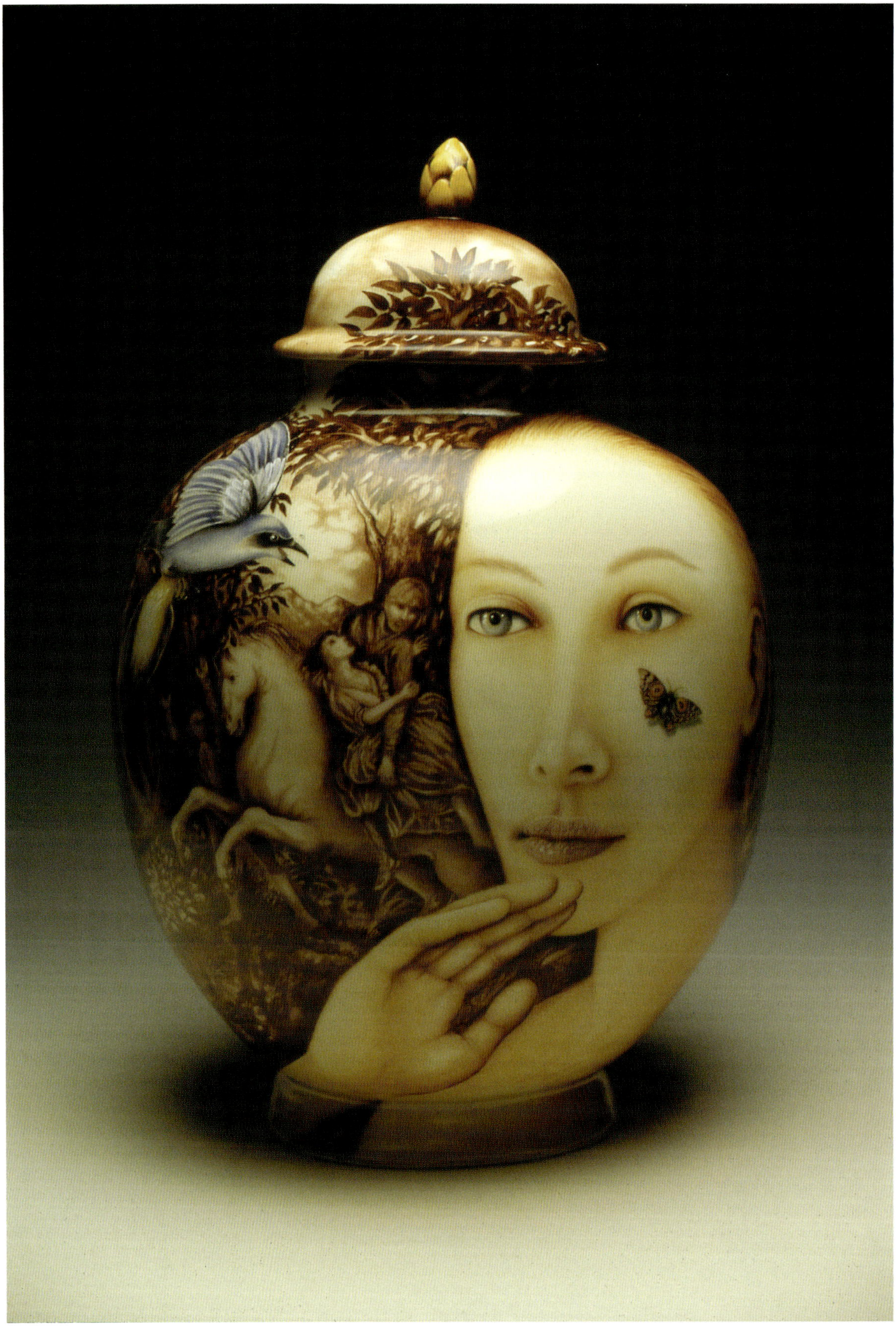

10

Abduction Jar, 1999 (front)

18 x 11 in.

Cast porcelain, china paint

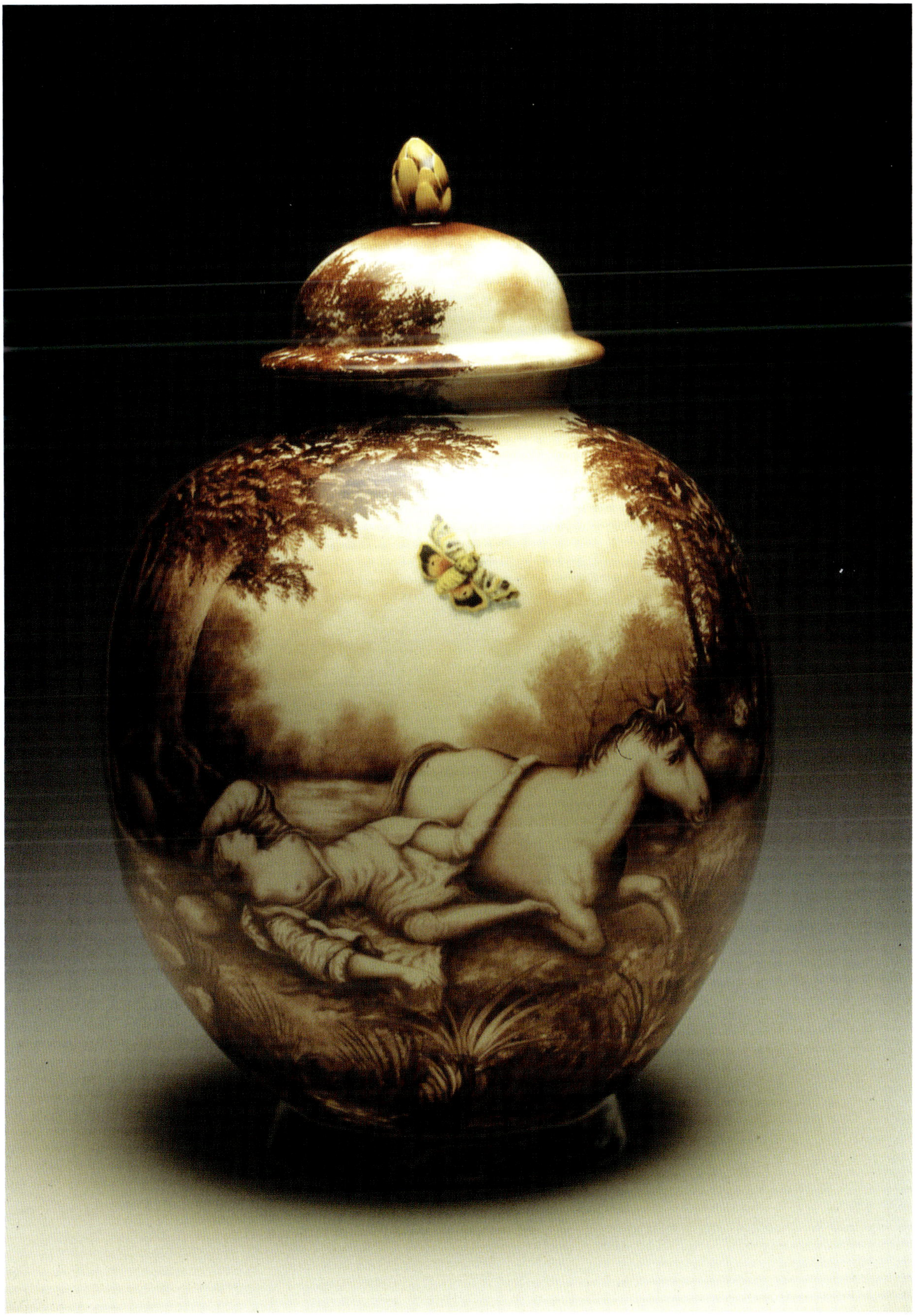

10

Abduction Jar, 1999 (back)

18 x 11 in.

Cast porcelain, china paint

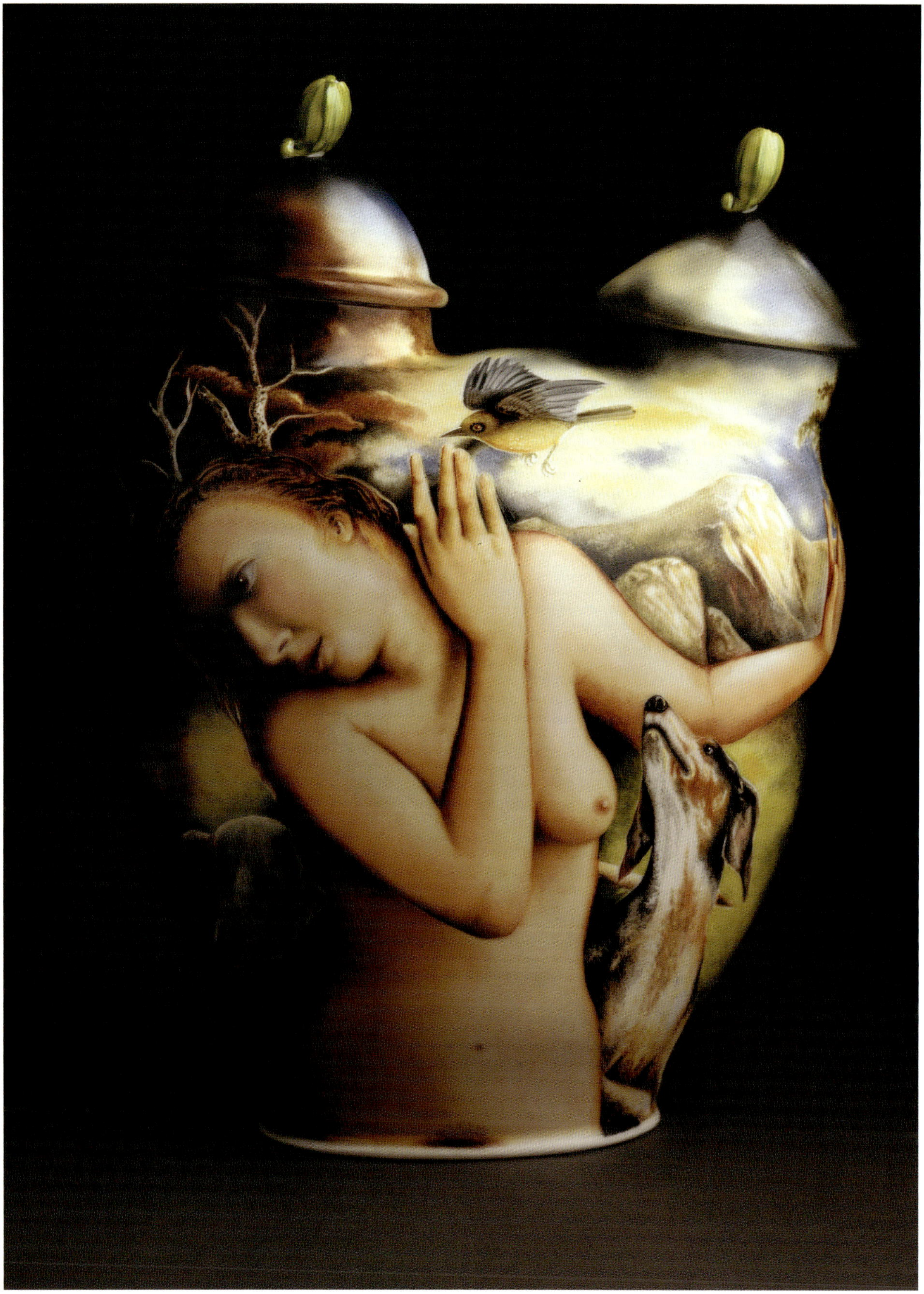

11

Fidelity, 2000 (front)

19 x 13½ in.

Cast porcelain, china paint

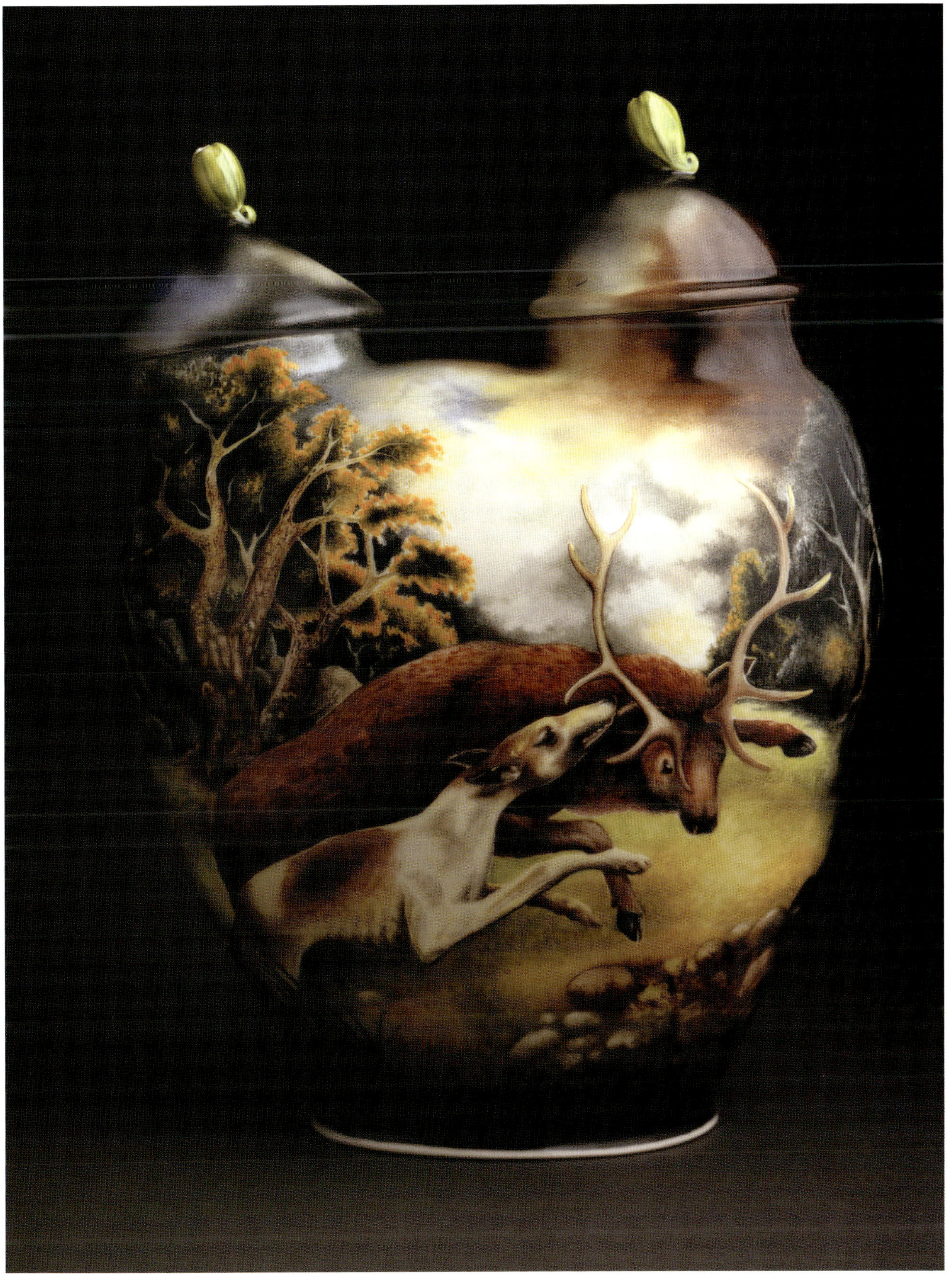

11

Fidelity, 2000 (back)

19 x 13½ in.

Cast porcelain, china paint

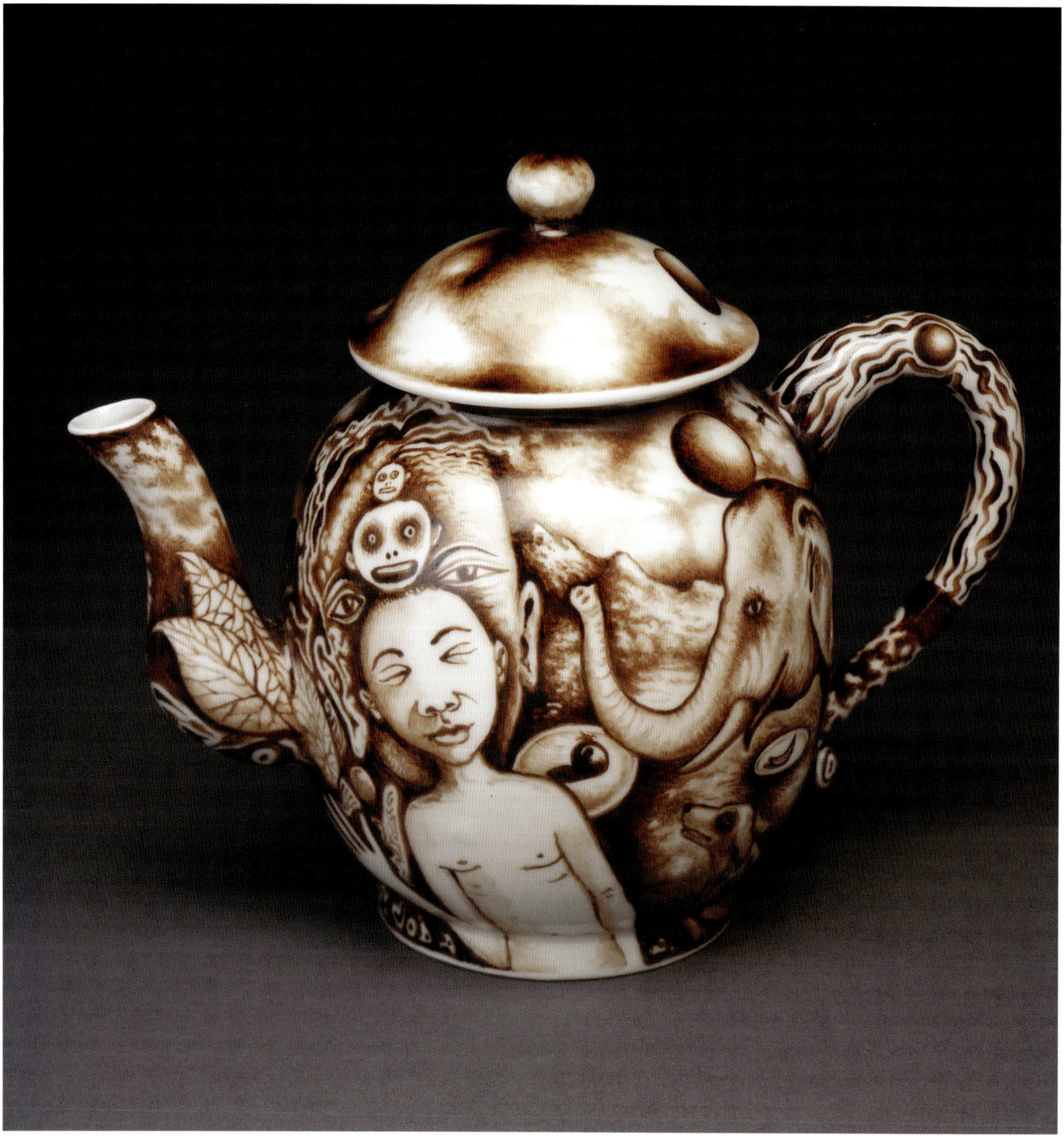

12

Untitled Teapot (Elephant), 2000

7 x 8 x 4½ in.

Cast porcelain, china paint

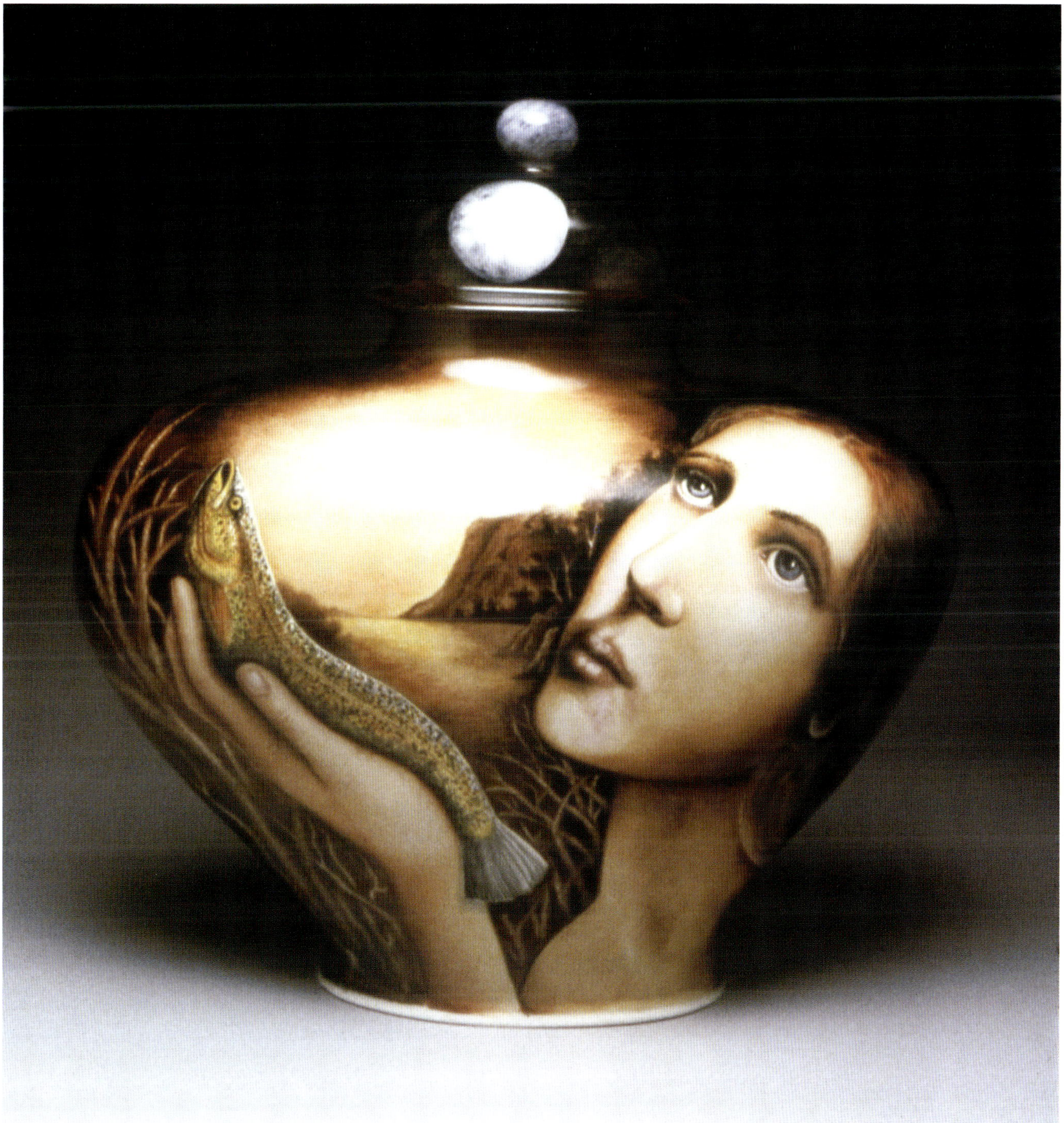

13

Water Clock, 2000

16½ x 16 in.

Cast porcelain, china paint

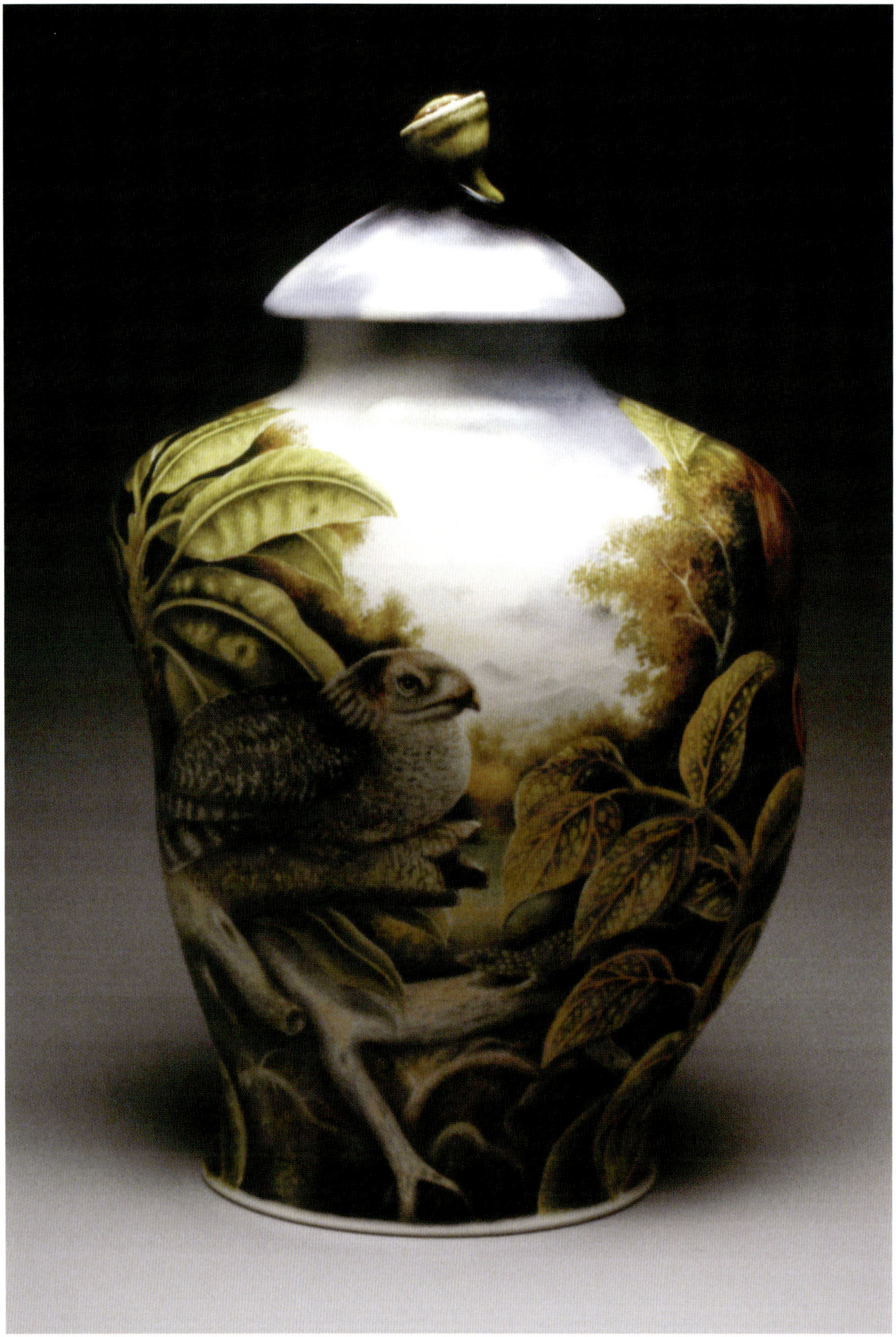

14

Perfume, 2000 (front)

14 x 12 x 7½ in.

Cast porcelain, china paint

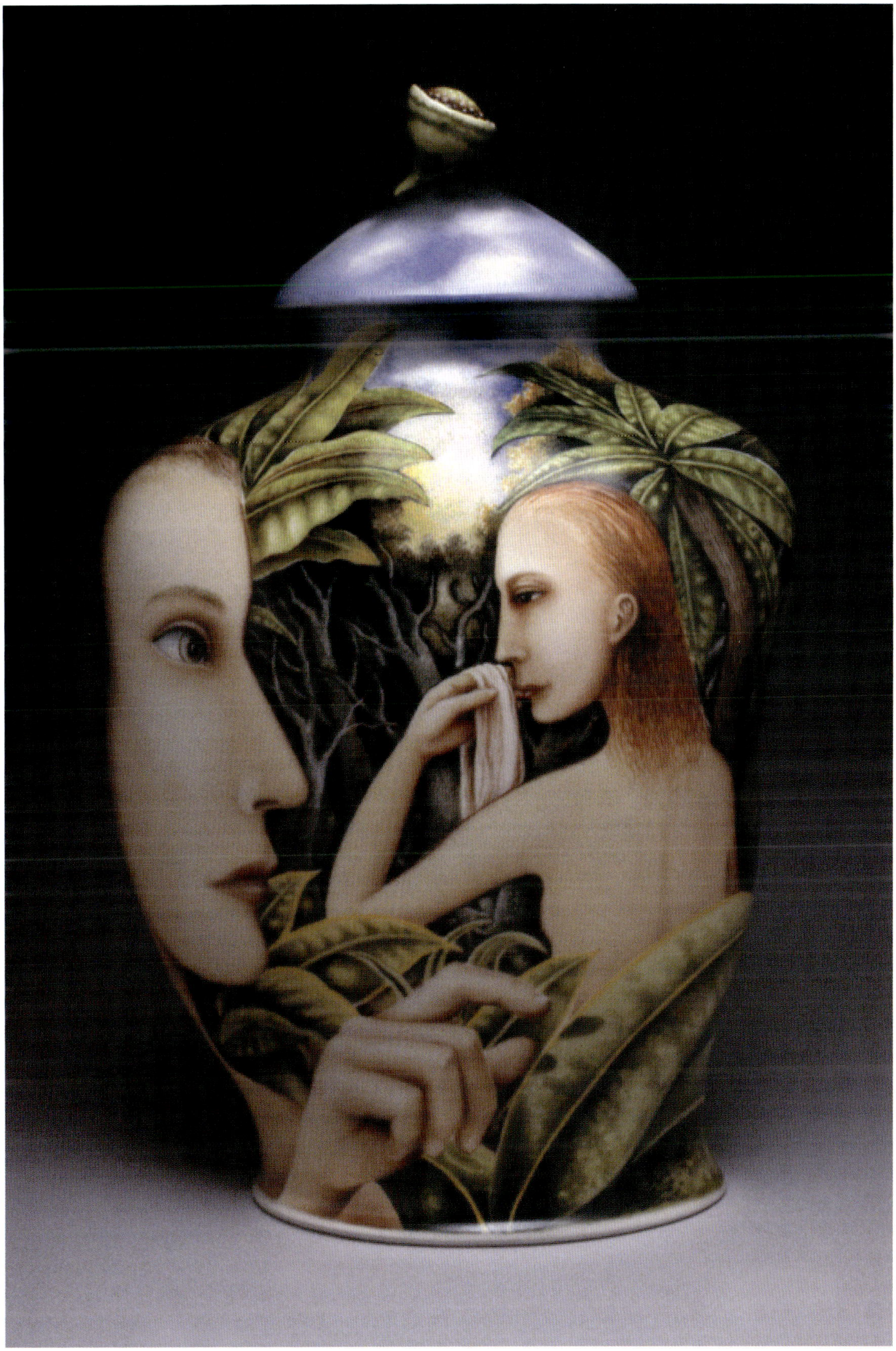

14

Perfume, 2000 (back)

14 x 12 x 7½ in.

Cast porcelain, china paint

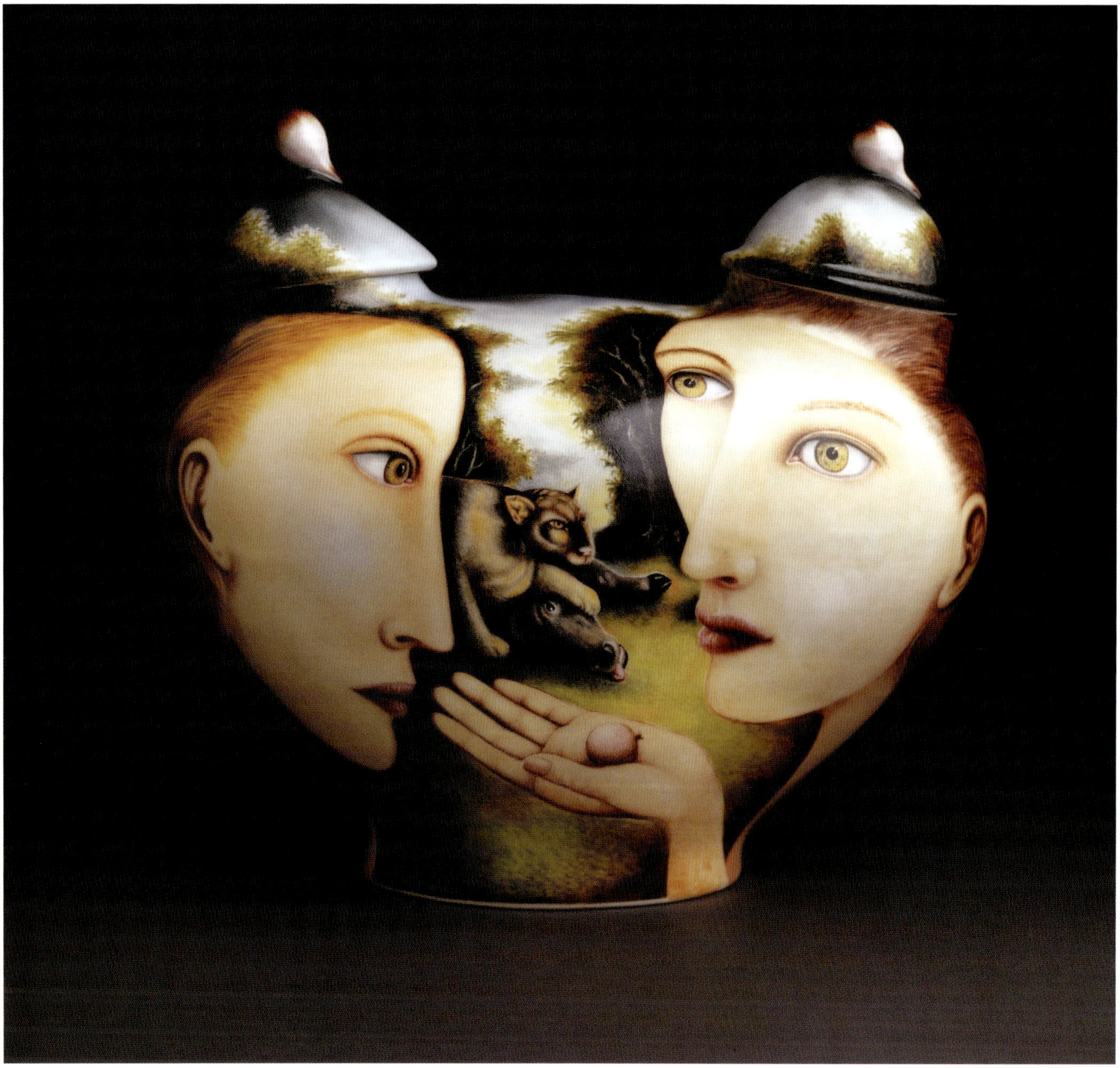

15

Small Fruit, 2000 (front)

15 x 16 in.

Cast porcelain, china paint

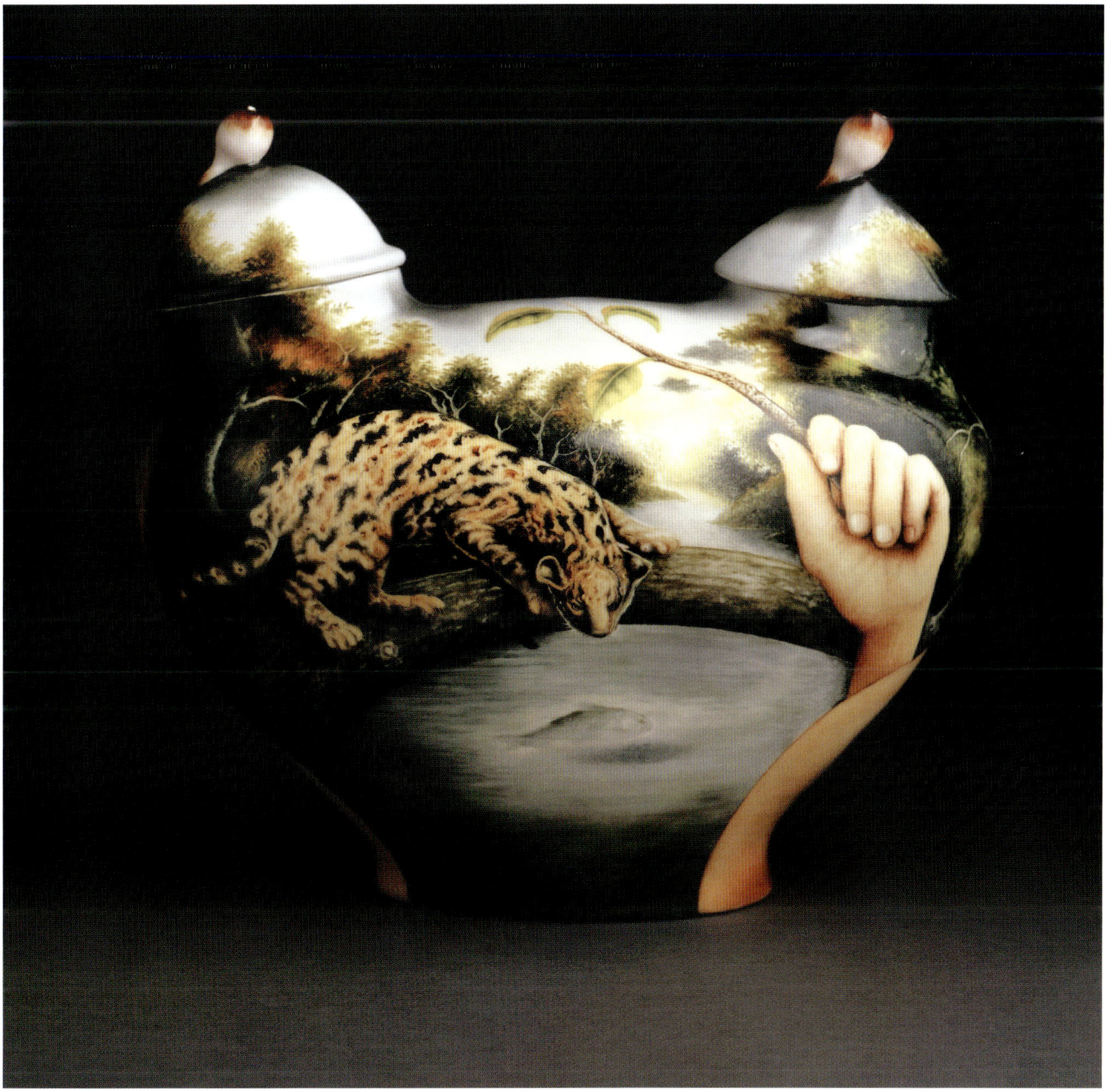

15

Small Fruit, 2000 (back)

15 x 16 in.

Cast porcelain, china paint

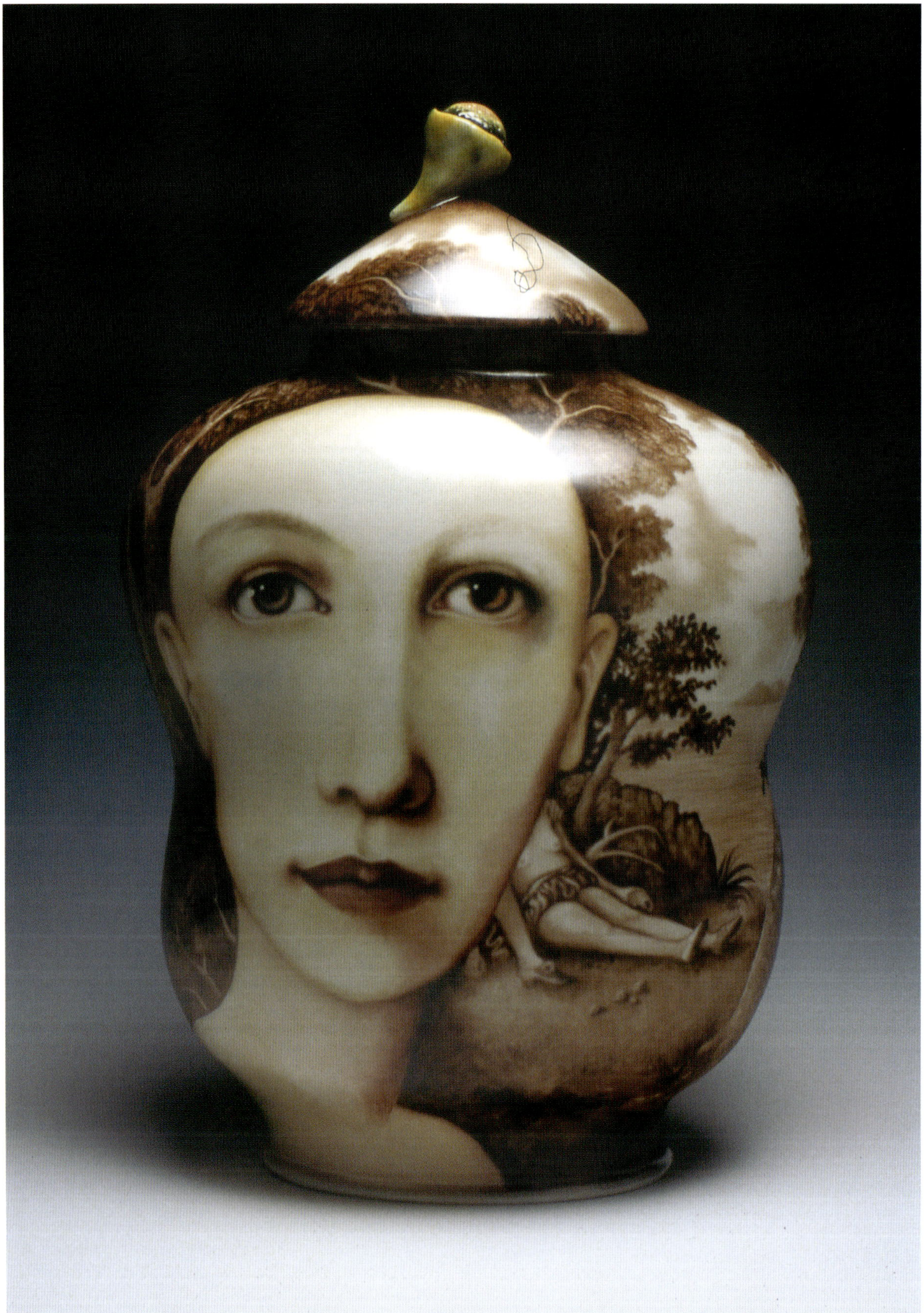

16

Bright Angel, 2001

17 x 12½ x 5½ in.

Cast porcelain, china paint

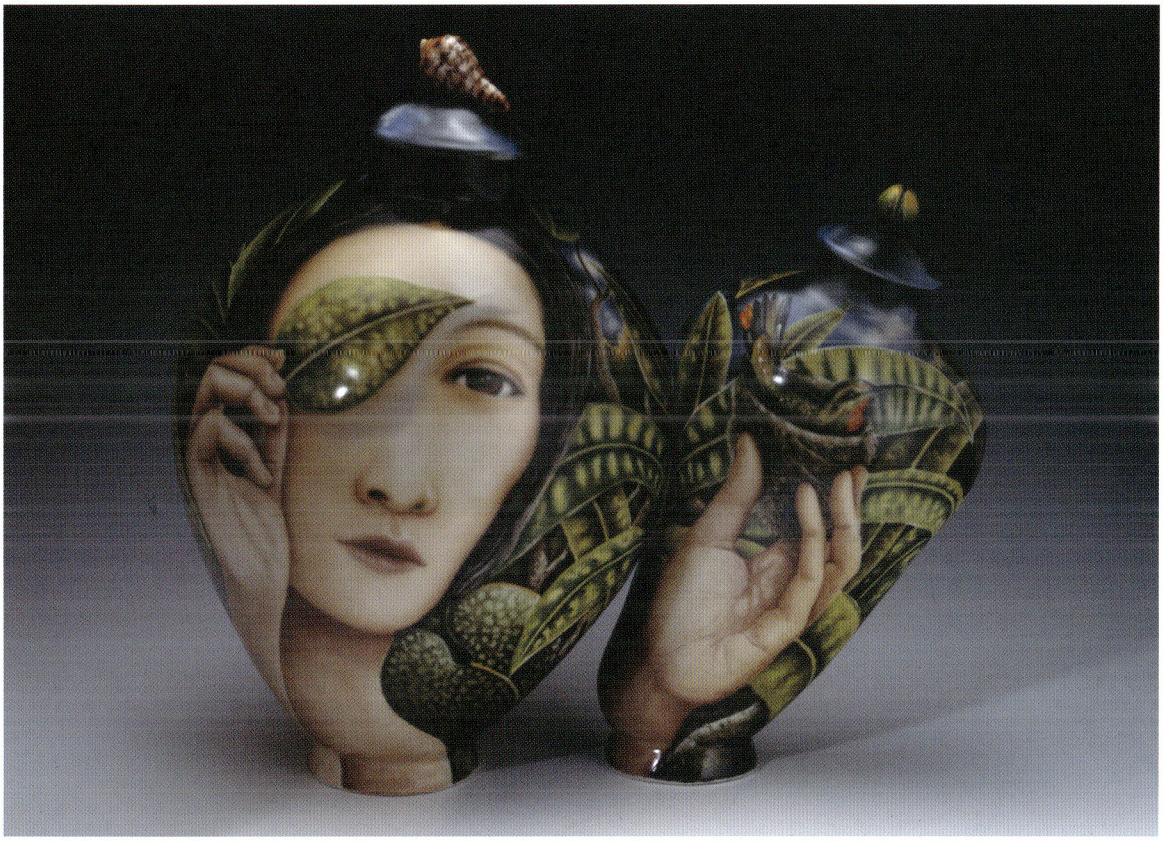

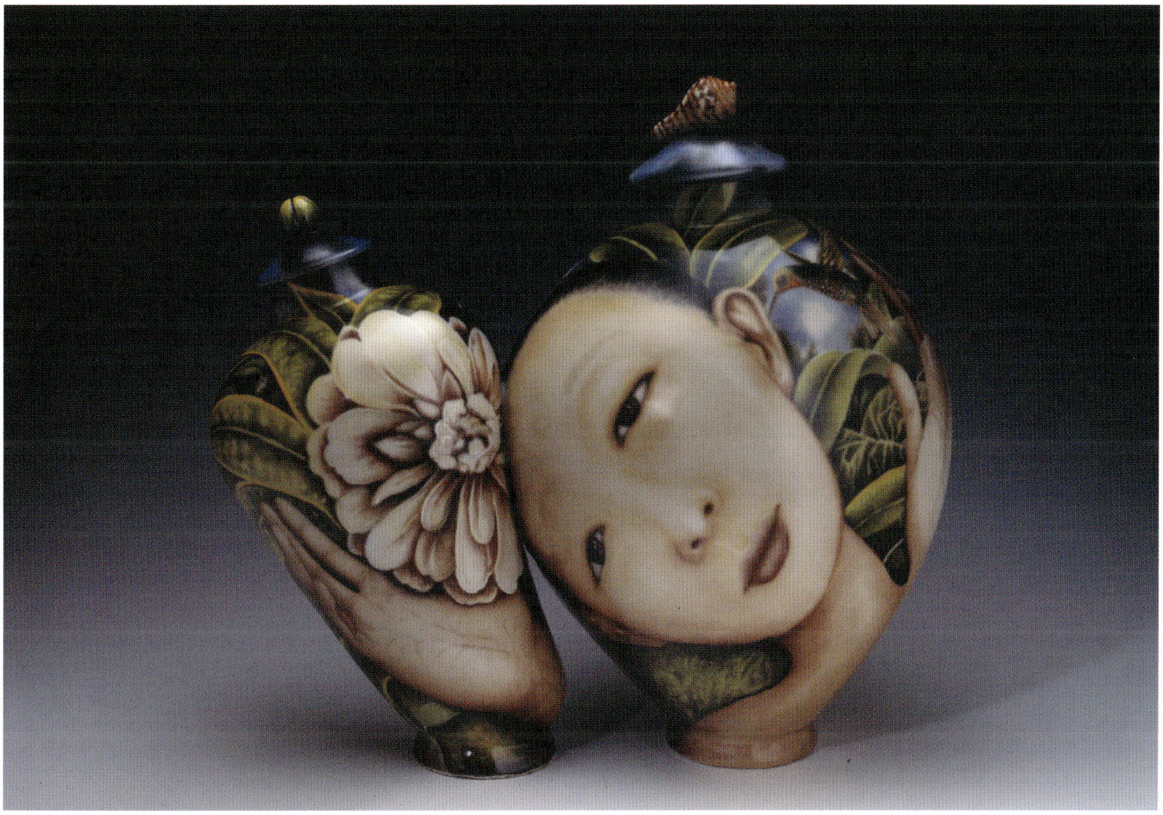

17

Semi-Conscious, 2002 (two views)

13 ½ x 14 in.

Cast porcelain, china paint

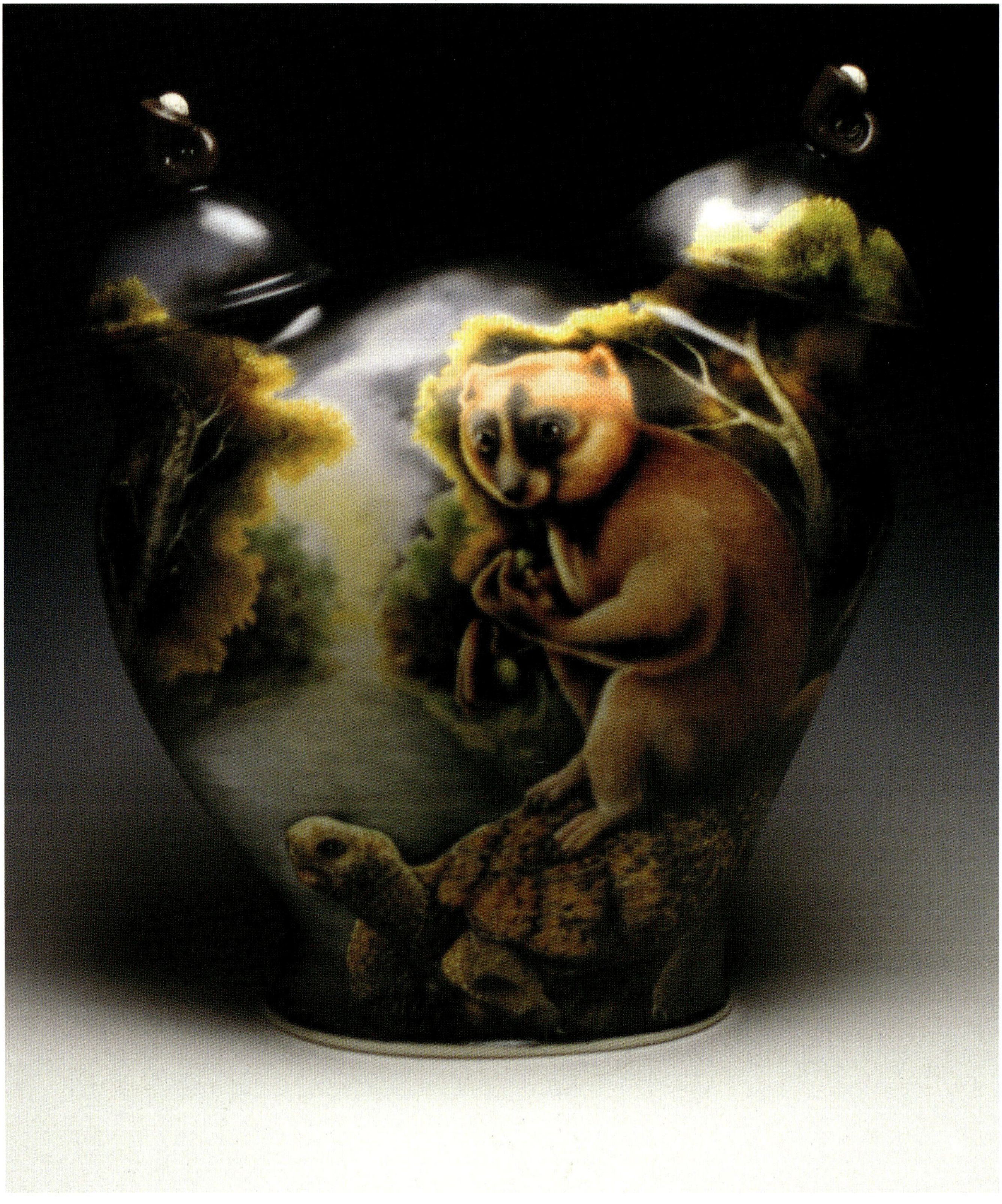

18

Orbit, 2002 (front)

17½ x 15 x 8½ in.

Cast porcelain, china paint

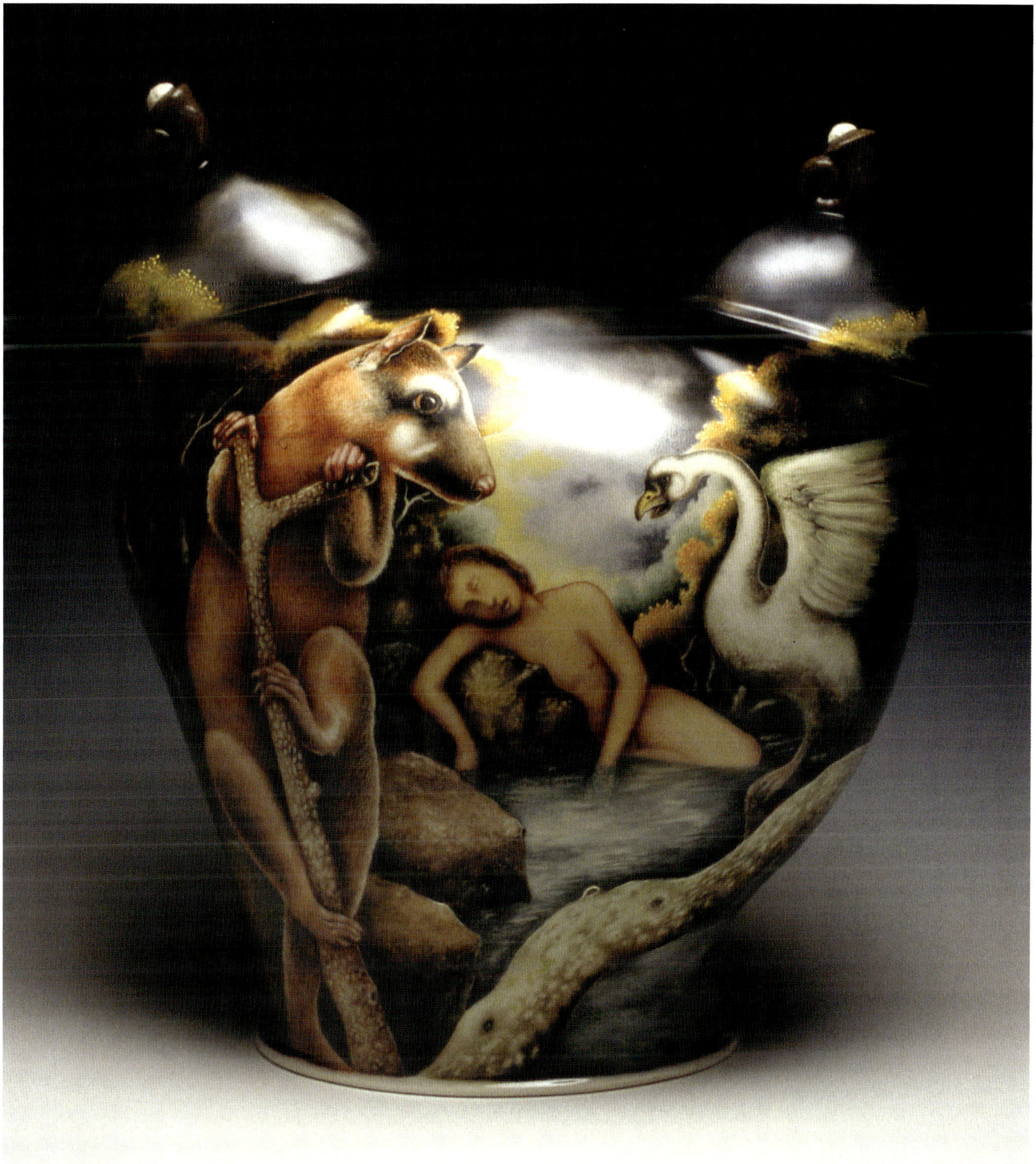

18

Orbit, 2002 (back)

17½ x 15 x 8½ in.

Cast porcelain, china paint

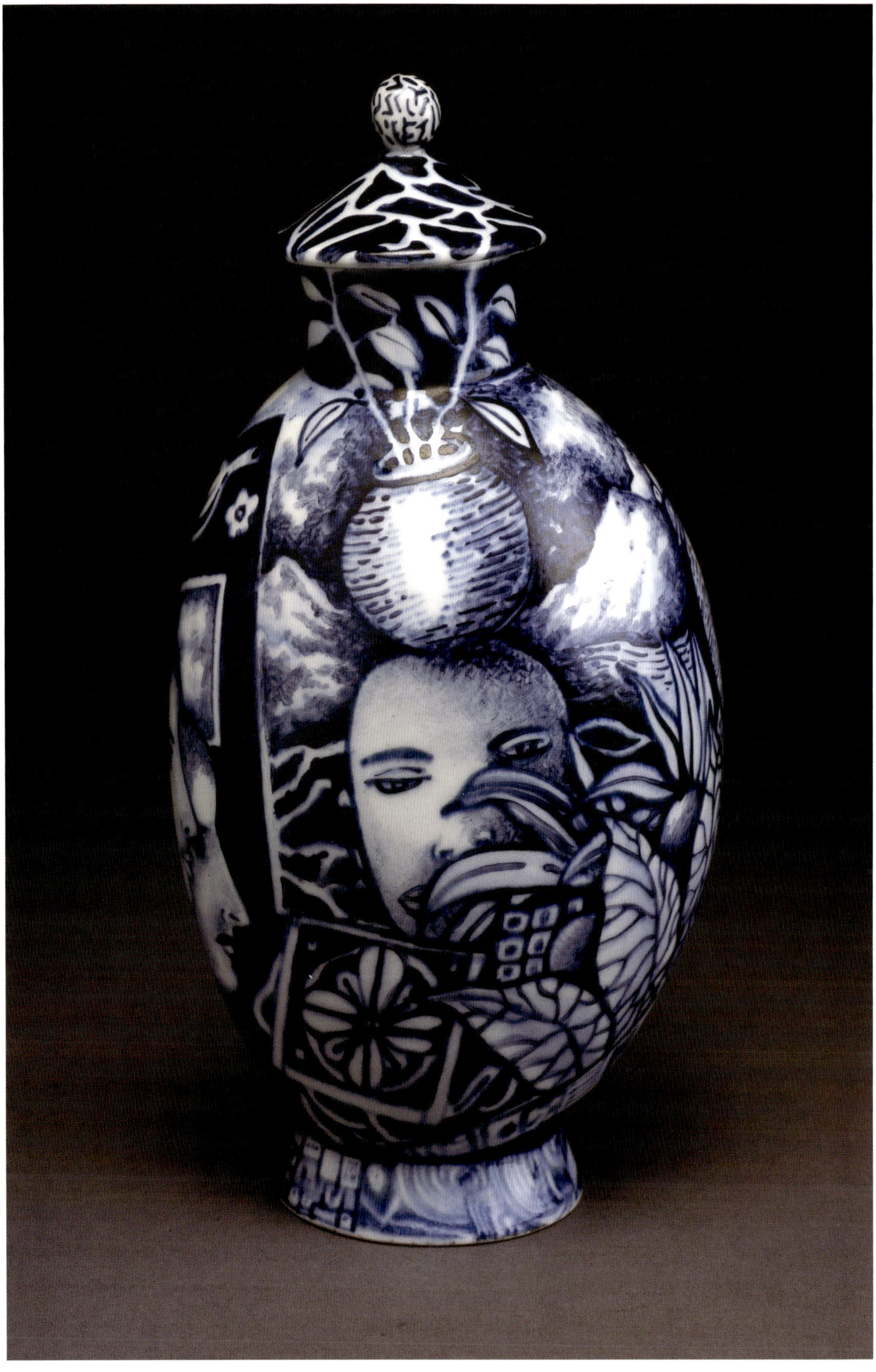

19

Blue-and-White Vase, 2003

9½ x 5 x 4 in.

Cast porcelain, cobalt decoration

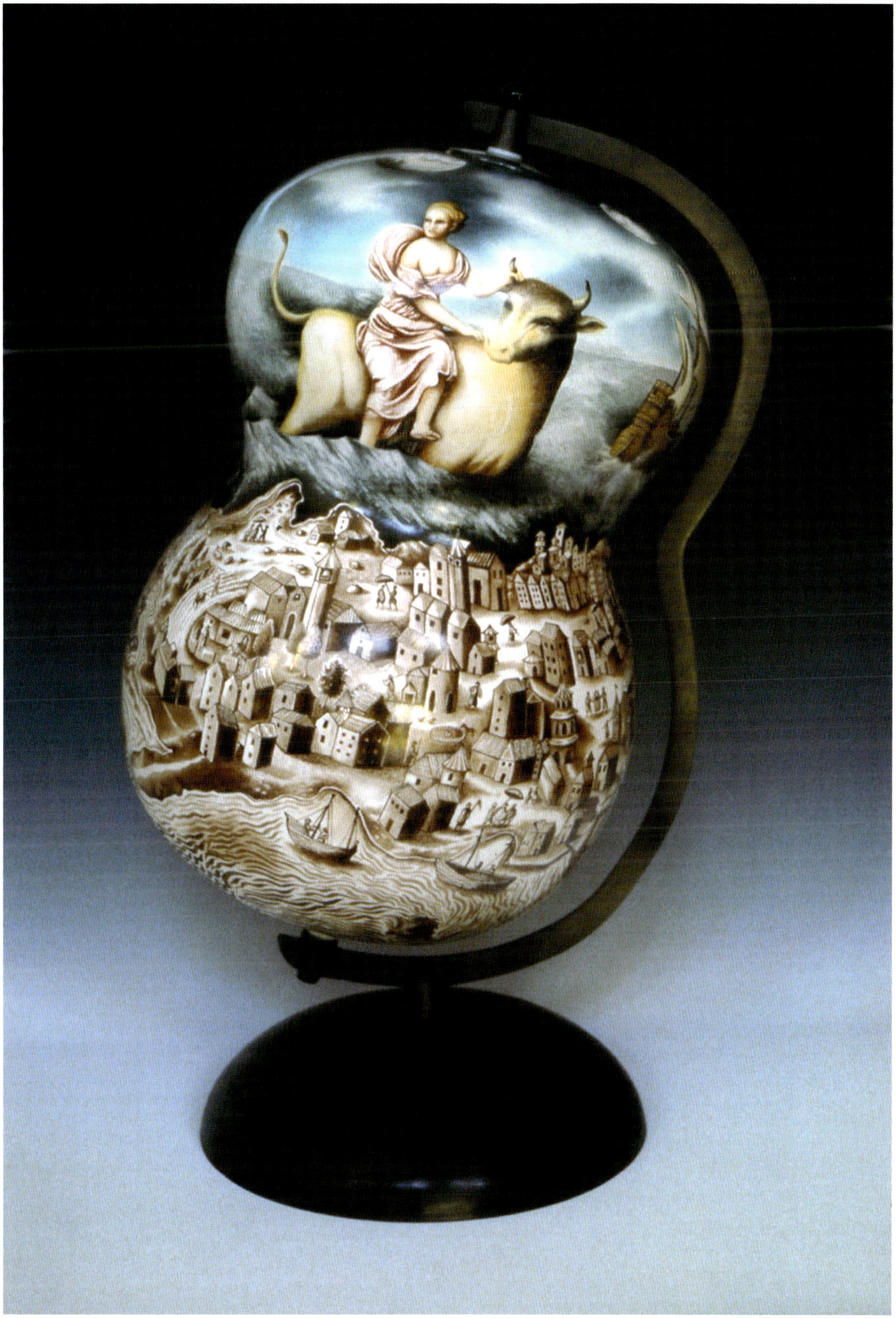

20

Europa, 2005

27 x 17 in.

Cast porcelain, china paint, bronze

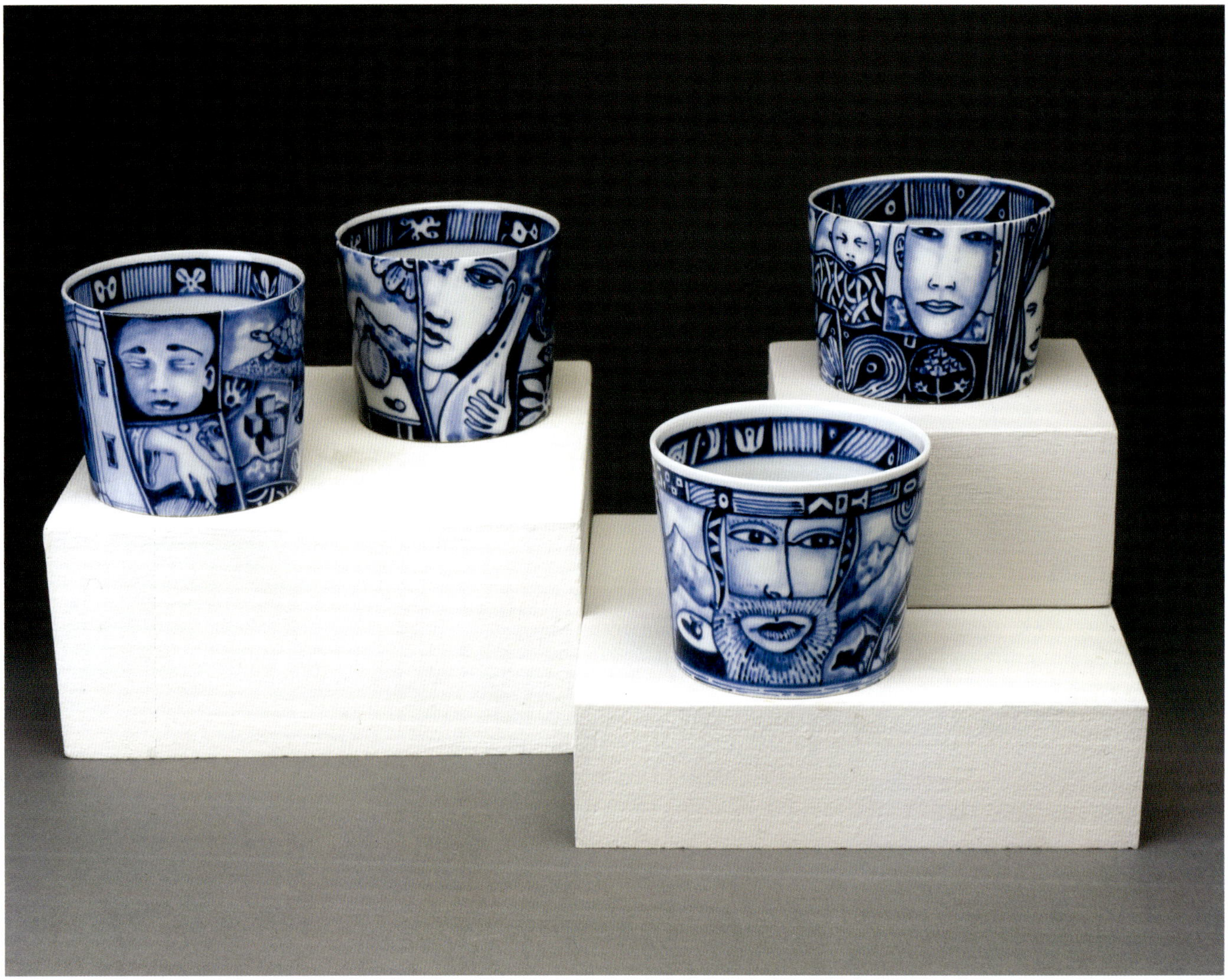

21

Set of Four Soba Choko Cups, 2006

Variable dimensions, approximately 3¼ x 3¾ in. diameter

Cast porcelain, cobalt decoration

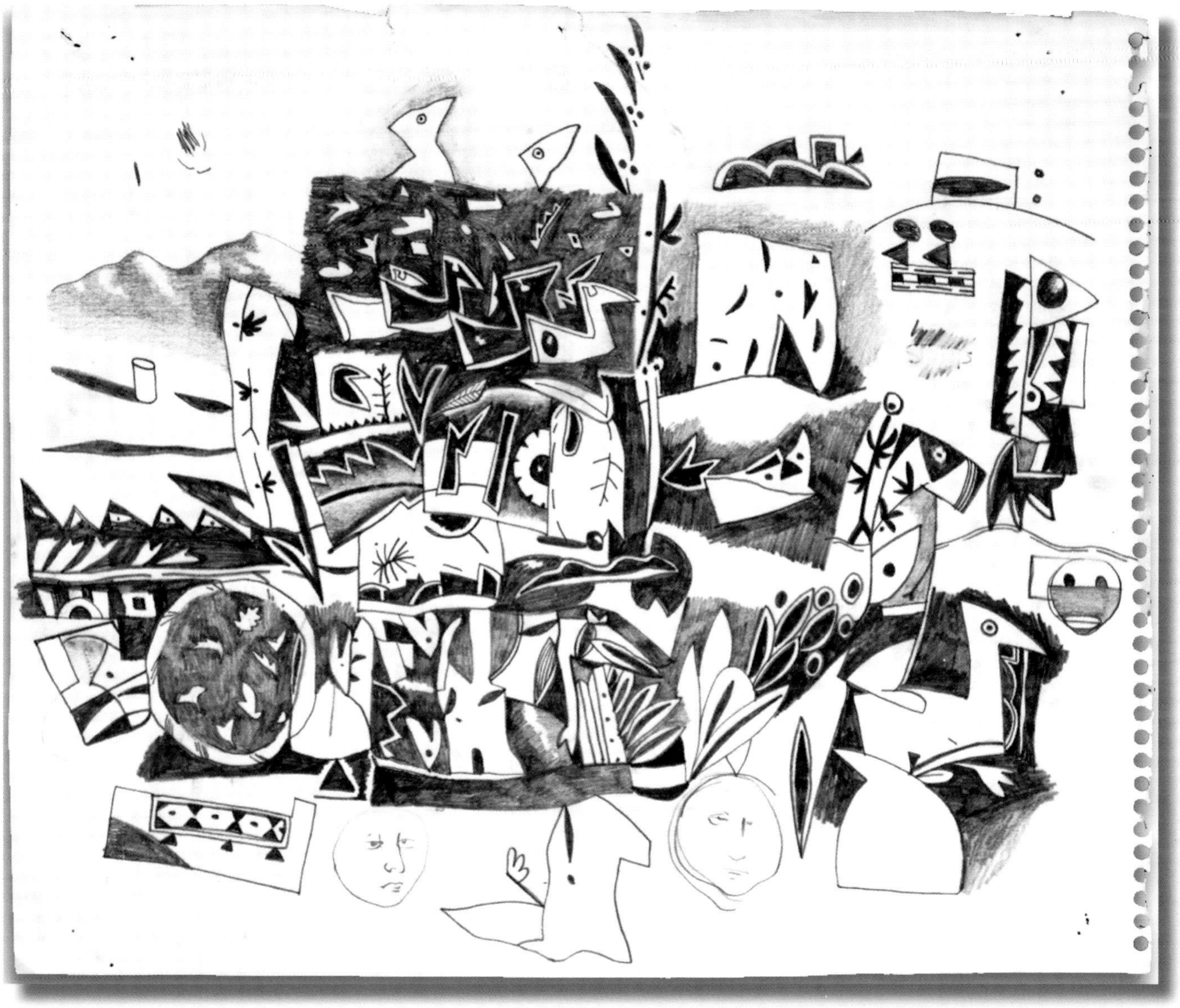

22

Untitled pencil drawing on paper, 1988
14 x 17 in.

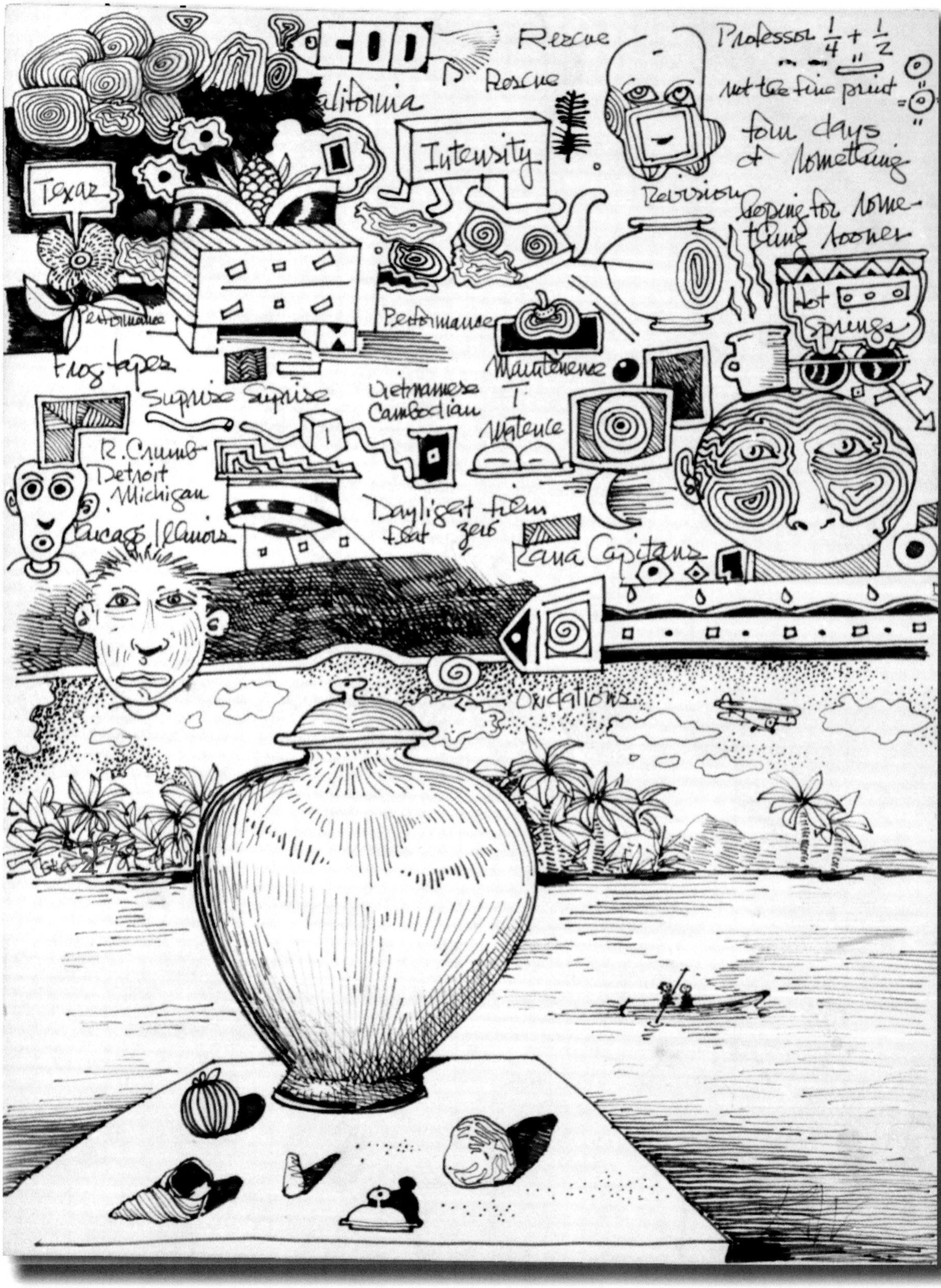

23

Untitled pencil drawing on paper, 1990

14 x 10½ in.

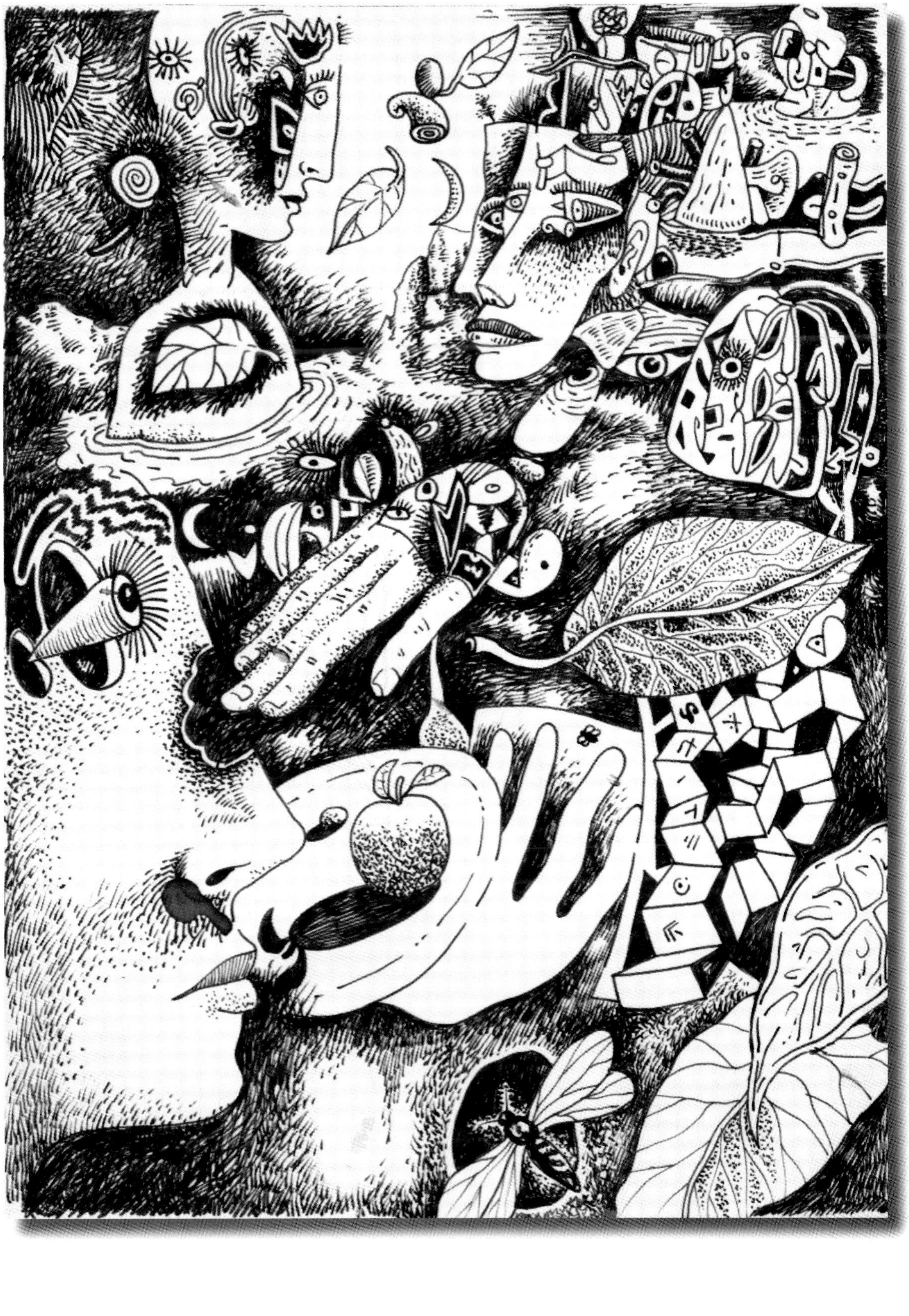

24

Untitled ink drawing on paper, 1996

11 x 8½ in.

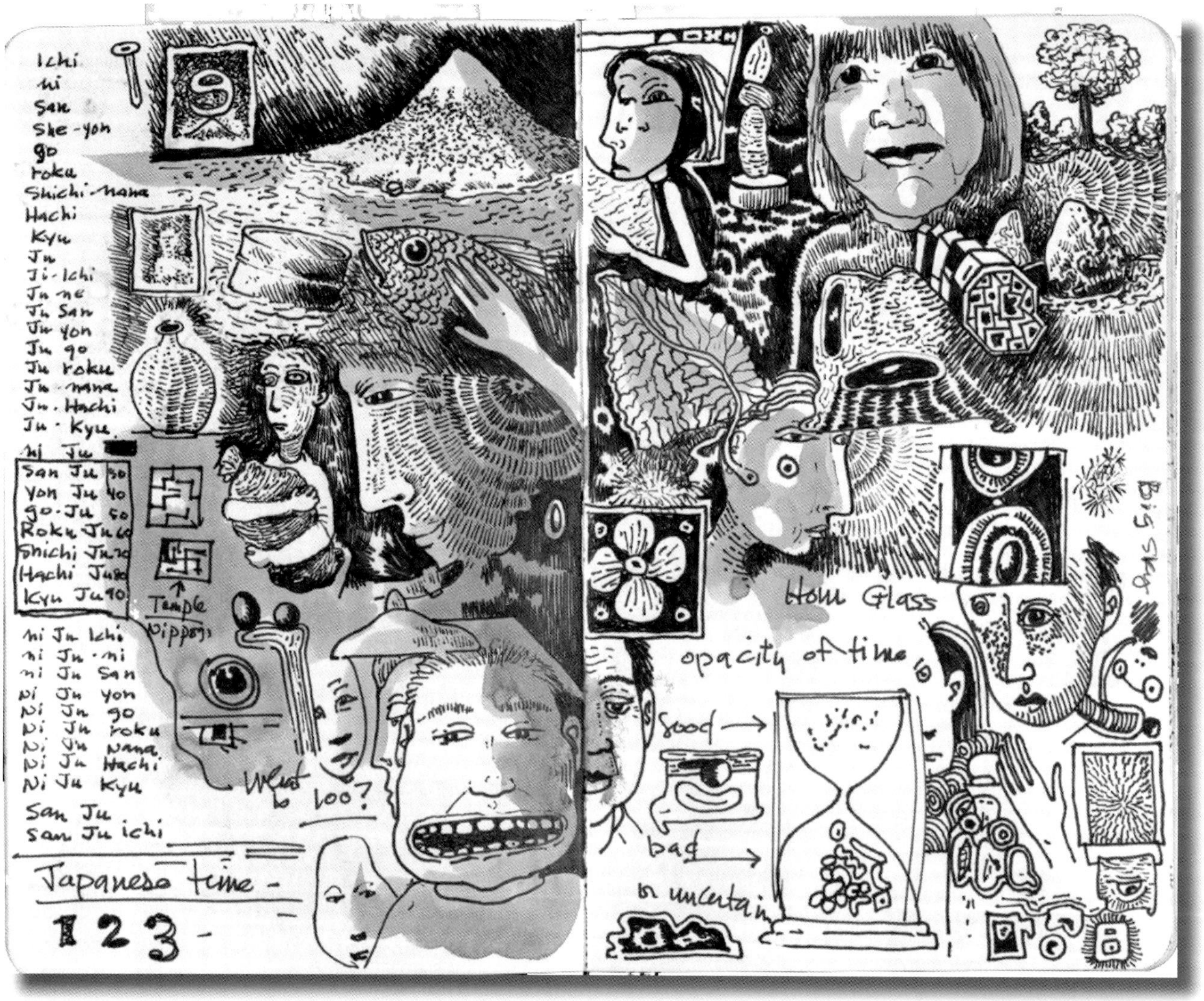

25a

Untitled ink drawing on paper (On plane ride to Calgary), 2003

(On plane ride to Calgary)

11 x 8½ in.

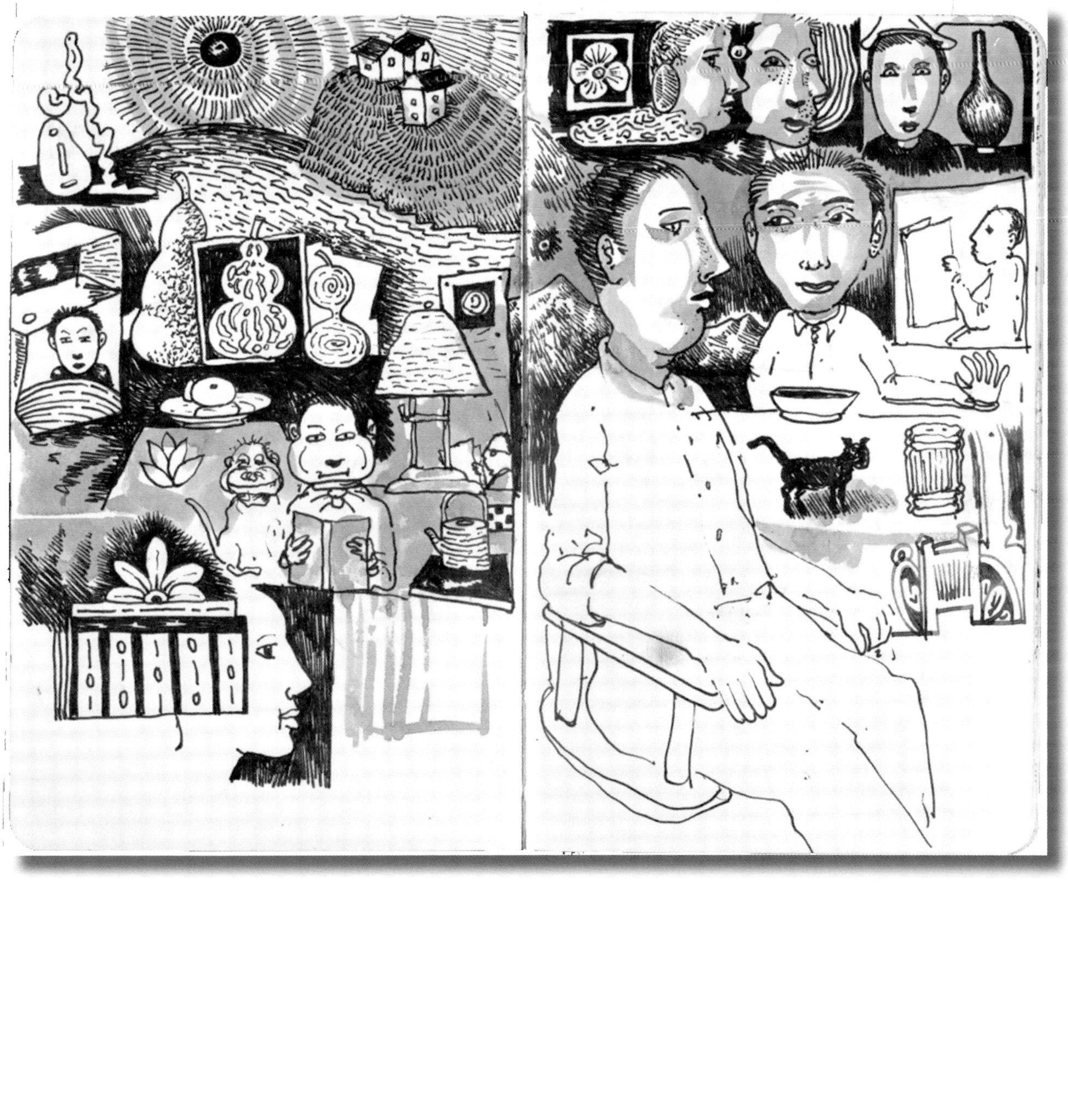

25b

Untitled ink drawing on paper, 2004
11 x 8½ in.

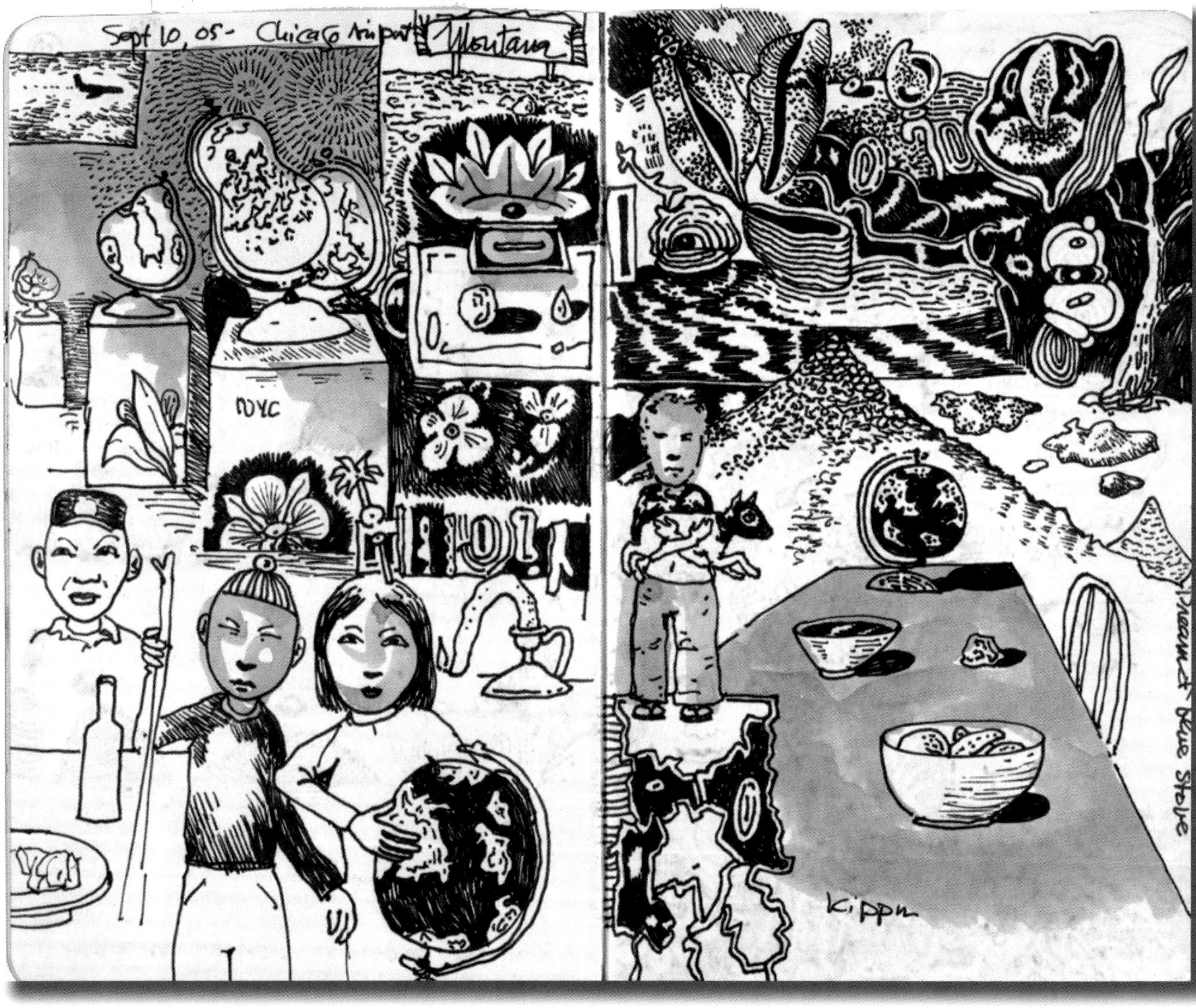

26

Untitled ink drawing on paper (On plane ride), 2005

11 x 8½ in.

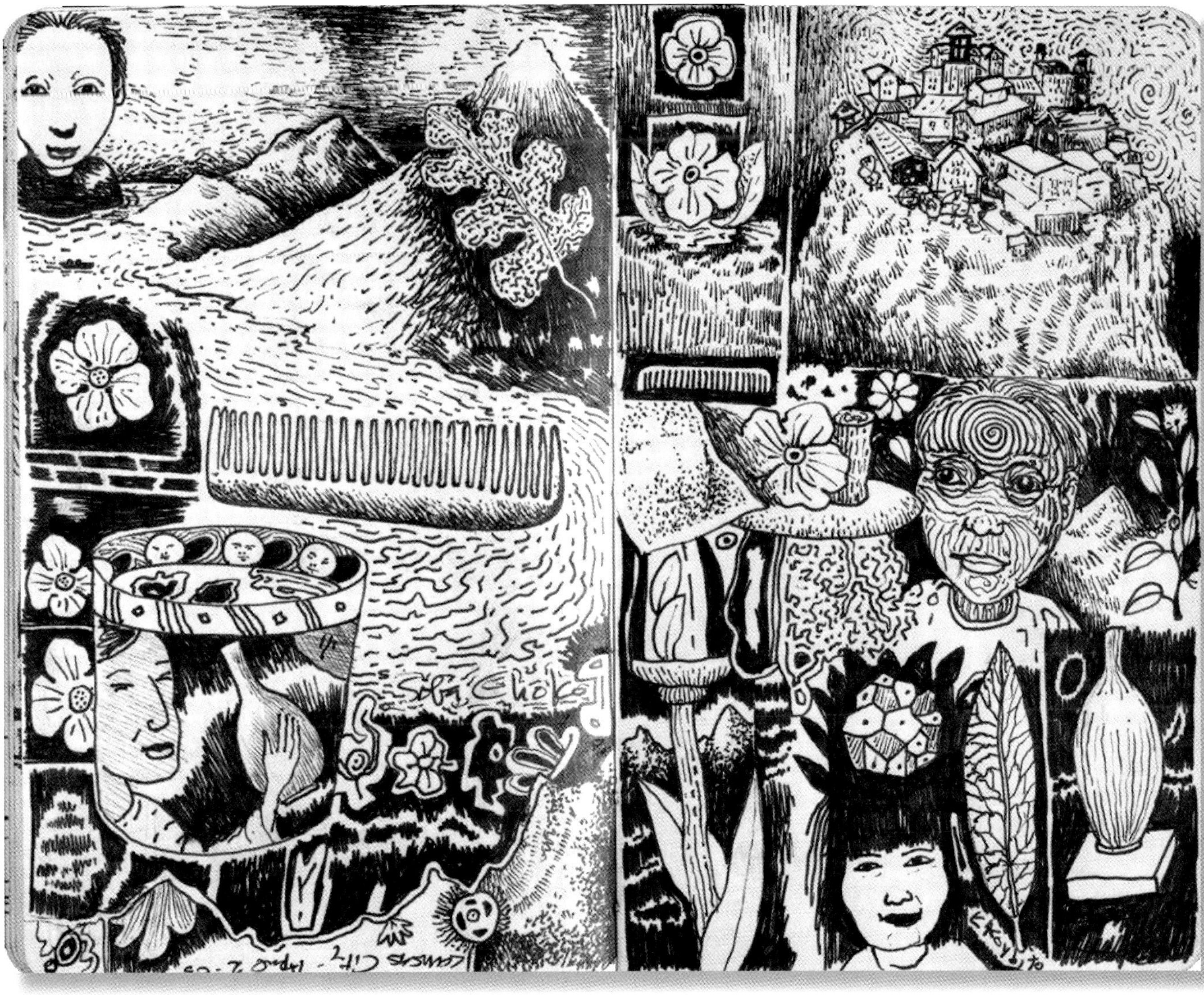

27

Untitled ink drawing on paper, 2005

12 x 9 in.

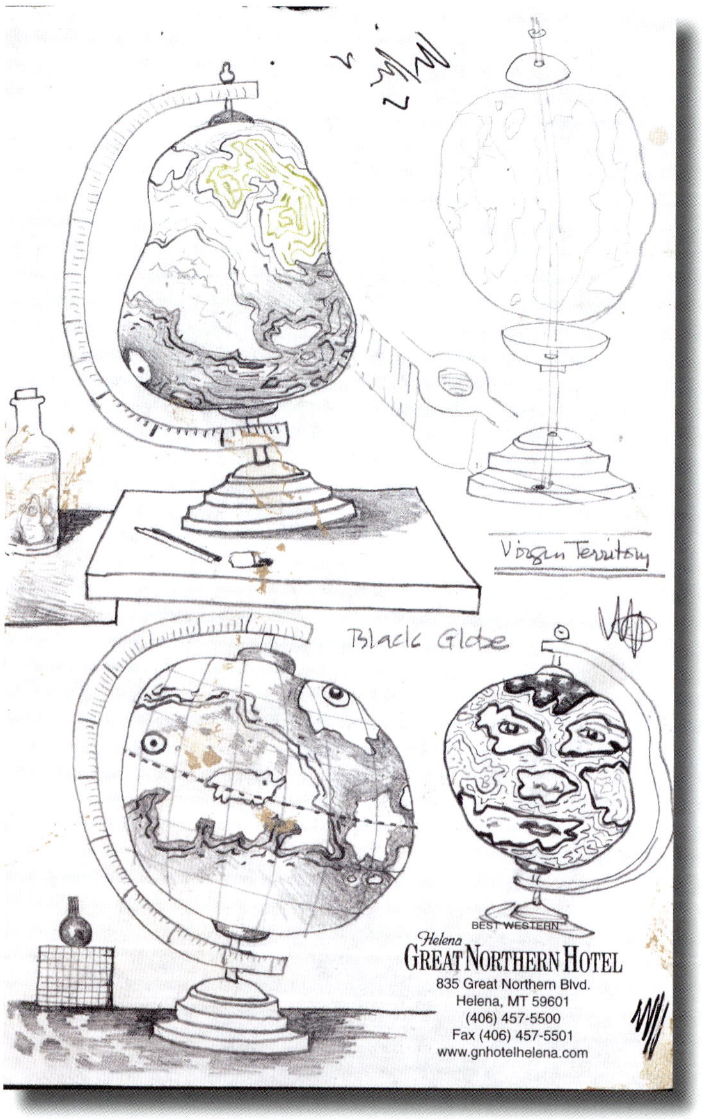

28

Untitled pencil drawing on paper (Made during Archie Bray Foundation board meeting), 2004

(Study for Globe Series)

8½ x 5½ in.

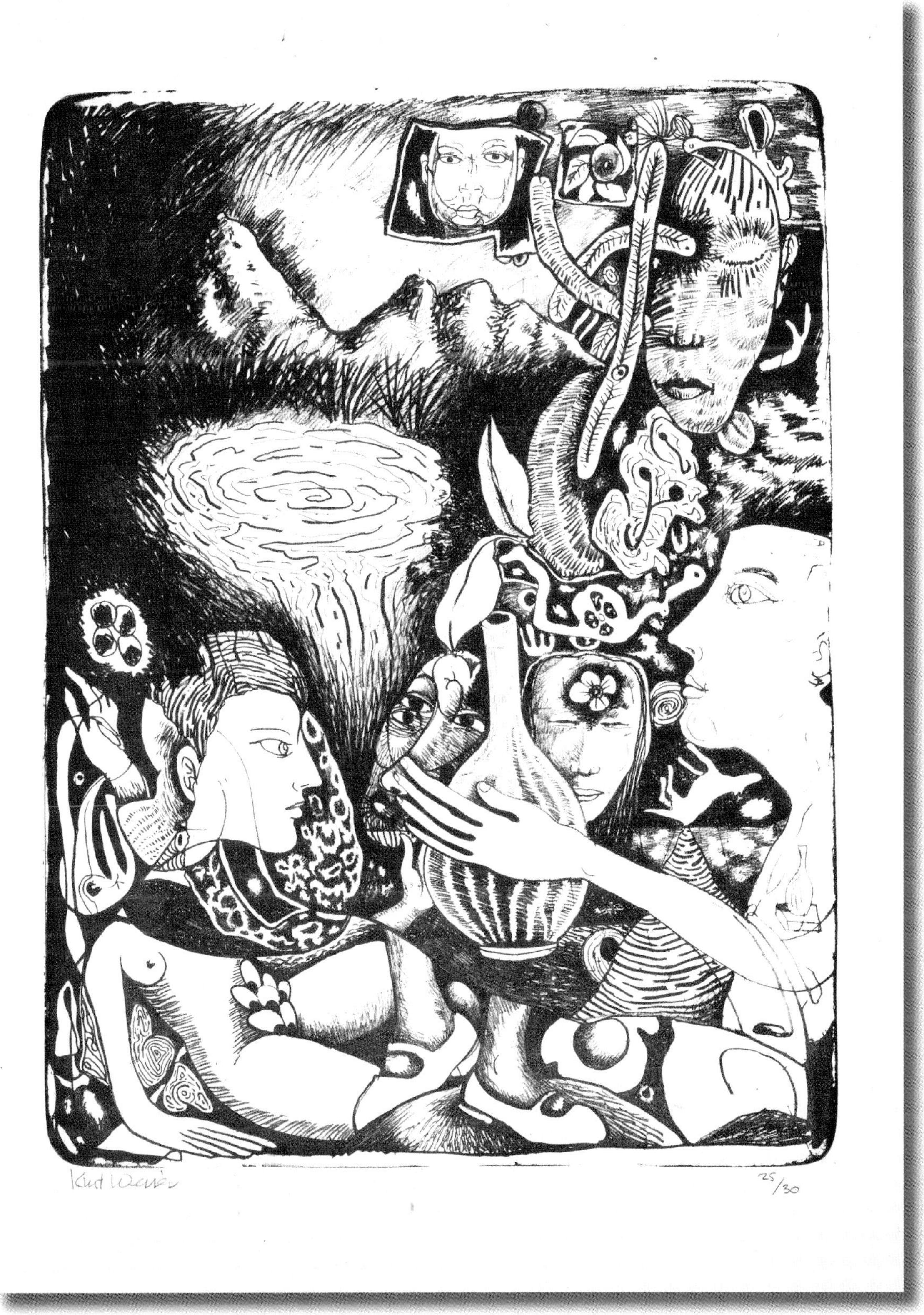

29

Untitled, 2006

Lithograph, 25/30

23½ x 17½ in.

Checklist of the Exhibition

Plate, 1977 (p. 19)
2½ x 20 in. diameter
Glazed stoneware
Collection of the Archie Bray Foundation for the Ceramic Arts

Plate, 1978 (p. 19)
2½ x 21¼ diameter
Glazed stoneware
Collection of Jeanne and William Porter

Teapot, ca. 1978 (p. 18)
10 x 6¾ x 7½ in.
Glazed porcelain
Collection of the Archie Bray Foundation for the Ceramic Arts

Raku Stirrup Jar, 1981 (p. 20)
20 x 16½ x 16½ in.
Glazed earthenware
Private Collection

Raku Toucan Jar, 1981 (p. 20)
15 x 12½ x 12½ in.
Glazed earthenware
Collection of Jeanne and William Porter

Plate, 1982 (pl. 1)
26½ in. diameter
Earthenware, residual salt-fired
Collection of the Archie Bray Foundation for the Ceramic Arts

Vase, 1984 (pl. 2)
17½ x 10 x 7¼ in.
Glazed stoneware
Collection of Barbara and Mike Jenson

Untitled Vase, 1984 (p. 22)
15 x 9½ x 5 in.
Glazed porcelain
Collection of Steven Korff and Marcia Van Wagner

Vessel with Stirrup Handle, 1984 (p. 21)
17 x 16 x 16 in.
Earthenware, residual salt-fired
Collection of the Arizona State University Art Museum
Gift of Anne and Sam Davis

Stepped Vase, ca. 1985 (p. 22)
16½ x 9 x 5½ in.
Glazed porcelain
Collection of the artist

Shino Crackle Cut Foot Vase, 1986 (pl. 3)
12 x 8 x 4 in.
Glazed stoneware
Collection of the artist

Blue & White Crackle Bowl, 1986 (p. 21)
5¾ x 19¾ x 19 in.
Glazed stoneware
Collection of Jeanne and William Porter

Leaf Plate, 1988 (p. 31)
4½ x 27 in. diameter
Glazed stoneware
Collection of Michael Gilbert & Catherine Shaner Gilbert

Tango, 1988 (p. 23)
22 x 10 x 5 in.
Glazed stoneware with porcelain inclusions
Collection of the artist

Vessel, 1988 (p. 23)
25½ x 10½ x 5 in.
Glazed stoneware with porcelain inclusions
Collection of the Archie Bray Foundation for the Ceramic Arts

Whispering Fish, 1991 (p. 33)
9¾ x 13 x 4½ in.
Cast porcelain, black slip
Collection of Craig Smith

Untitled Teapot (Scissors in Hand), 1991 (p. 33)
11 x 13 x 4 in.
Cast porcelain, black slip
Courtesy of the Kamm Teapot Foundation

The Bird Merchant, 1991 (p. 34)
11 x 12½ x 4½ in.
Cast porcelain, china paint
Courtesy of the Kamm Teapot Foundation

Lidded Jar, 1992 (pl. 5)
17 x 10 x 6 in.
Cast Porcelain, china paint
Collection of Stephen & Pamela Hootkin

Iguana, 1992 (pl. 4)
17¼ x 9 x 5 in.
Cast porcelain, china paint
Collection of the Arizona State University Art Museum, Gift of the Museum Store

Bug Lesson, 1994 (pl. 6)
17 x 12 in.
Cast porcelain, china paint
Collection of Jerome Shaw

Sleeper, 1994 (pl. 7)
20 x 12 x 6½ in.
Cast porcelain, china paint
Collection of Dr. and Mrs. George Andros

Marbleized Teapot with Hand, 1996 (pl. 8)
11¼ x 14¼ x 5 in.
Cast porcelain, china paint
Renwick Gallery of the Smithsonian American Art Museum, museum purchase through the Howard Kottler Endowment for Ceramic Art

Confidential, 1997 (p. 30)
18 x 12 in.
Cast porcelain, china paint
Collection of Jim and Vicki Mercer

Red Leaf, 1998 (p. 9)
18½ x 10½ x 7 in.
Cast porcelain, china paint
Collection of Roberta and Bruce Hammer

Abduction Jar, 1999 (pl. 10)
18 x 11 in.
Cast porcelain, china paint
Collection of Barbara and Donald Tober

Small Fruit, 2000 (pl. 15)
15½ x 16 in.
Cast porcelain, china paint
Collection of John Hoffman

Untitled Teapot (Elephant), 2000 (pl. 12)
7 x 8 x 4½ in.
Cast porcelain, china paint
Collection of the Arizona State University Art Museum, Museum Purchase

Water Clock, 2000 (pl. 13)
16½ x 16 in.
Cast porcelain, china paint
Sprint Nextel Art Foundation

Outdoor Life, 2000 (p. 35)
12 x 12 in.
Cast porcelain, china paint
Collection of Sandra Schwalb

Fidelity, 2000 (pl. 11)
19 x 13½ in.
Cast porcelain, china paint
Collection of Barbara Wild

Perfume, 2000 (pl. 14)
14 x 12 x 7½ in.
Cast porcelain, china paint
Collection of Dr. and Mrs. Youk Lee

Bright Angel, 2001 (pl. 16)
17 x 12½ x 5½ in.
Cast porcelain, china paint
Collection of Saul E. Levi

Orbit, 2002 (pl. 18)
17½ x 15 x 8½ in.
Cast porcelain, china paint
Collection of Pam Smith

Night Harvest, 2002 (p. 39)
17¾ x 12½ x 9 in.
Cast porcelain, china paint
2002.13A, B. Collection of the Newark Museum, Purchase 2002 membership Endowment Fund, Eleanor S, Upton Bequest Fund and Franklin Conklin Memorial Fund

Semi Conscious, 2002 (pl. 17)
Cast porcelain, china paint
13½ x 14 in.
Collection of Marvin Gerstein

Blue-and-White Vase, 2003 (pl. 19)
9½ x 5 x 4 in.
Cast porcelain, cobalt decoration
Collection of the Arizona State University Art Museum, Diane & Sandy Besser Collection

Europa, 2005 (p. 39, pl. 20)
27 x 17 in.
Cast porcelain, china paint, bronze
Collection of Sara and David Lieberman

Many Many, 2005 (p. 36)
25 x 21 in.
Cast porcelain, china paint, bronze
Collection of Jeanne and William Porter

Set of Four Soba Choko Cups, 2006 (pl. 21)
Variable dimensions, approximately 3¼ x 3¾ x 3¾
Cast porcelain, cobalt decoration
Collection of the artist

Prints and Drawings

Untitled pencil drawing on paper, 1988 (pl. 22)
14 x 17 in.
Courtesy of the artist

Untitled pencil drawing on paper, 1990 (pl. 23)
14 x 10 ½ in.
Courtesy of the artist

Untitled ink drawing on paper, 1996 (pl. 24)
11 x 8 ½ in.
Courtesy of the artist

Untitled ink drawing on paper (on plane ride to Calgary), 2003 (pl. 25a)
11 x 8½ in.,
Courtesy of the artist

Untitled ink drawing on paper (on plane ride), 2004 (pl. 25b)
11 x 8½ in.,
Courtesy of the artist

Untitled ink drawing on paper (on plane ride), 2005 (pl. 26)
11 x 8½ in.,
Courtesy of the artist

Untitled ink drawing on paper, 2005 (pl. 27)
12 x 9 in.
Courtesy of the artist

Untitled pencil drawing on paper (Made during Archie Bray Foundation board meeting), 2004
(study for *Globe Series*) (pl. 28)
8½ x 5½ in.
Courtesy of the artist

Untitled, 2006 (pl. 29)
23½ x 17½ in., unframed
Lithograph, 25/30
Printed with Joe Segura Press, Mesa, Arizona
Collection of the Arizona State University Art Museum, Museum Purchase, Gift of the artist

Chronology

1950
Born July 17th in Lansing, Michigan, the oldest of five children. His father Douglas was a publicist for Michigan State University; his mother Elizabeth was a secretary for the Michigan Highway Department. Raised in a Catholic home, Kurt is fascinated by religious iconography.

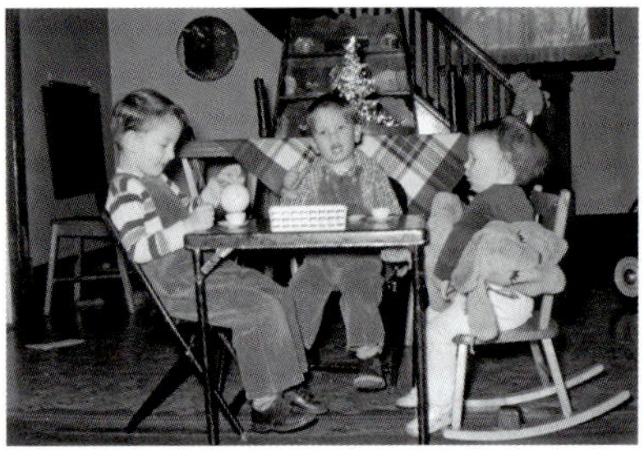

Kurt, on left, pours tea for brother Tom, center, and sister Marybeth in the family home, 1955.

1956
A local clay deposit found close to the home provided the first experience in modeling figures and making pipes, which were fired in the family's fireplace.

1960
While in 5th grade, creates a ceramic mural depicting nature and animal forms for the Red Cedar Elementary School that is still installed in their lobby.

1967
Kurt, struggling within a rigid academic environment, drops out of high school in the 9th grade. Works at several jobs, including playing drums for a college rock and roll band. His mother, recognizing his precocious talents, suggests enrolling at the Interlochen Arts Academy in Interlochen, Michigan. Kurt lives on campus and is immersed in the arts curriculum. His ceramics teacher Jean Parsons, trained under Maija Grotell at Cranbrook Academy, is a proponent of modernist Scandinavian design.

1969
Graduates from Interlochen Arts Academy and enrolls briefly at the Detroit Society of Arts & Crafts. Transfers to the Kansas City Art Institute, in Kansas City, Missouri. Studies ceramics with Ken Ferguson, Victor Babu and Jacquie Rice. Focuses on making functional stoneware and raku pottery. Sells work at local craft fairs, notably at the Ann Arbor Street Fair, to help pay for living expenses.

1970
Meets future wife Christy Lasater, who is enrolled in the graphic design program.

1972
Completes studies at Kansas City Art Institute, graduating with a B.F.A. Fellow. Classmates include Andrea and John Gill, Rick Hensley, Richard Notkin, Donna Polseno, Stanley Welsh, and Allan Winkler. Ken Ferguson recommends Weiser to Oregon potter Tom Coleman, who is searching for a ceramics instructor at the Portland Museum Art School. Kurt heads west for the first time; Christy has one year left to complete her studies at KCAI.

1973
Trades jobs with potter Pat Horsley and goes on to teach at Marylhurst College, Marylhurst, Oregon. Christy joins Kurt and are soon married. Growing weary of the Oregon environment and the over crowded pottery scene, ponders returning to school for graduate studies.

1974
Enrolls in the graduate ceramics program at the University of Michigan, Ann Arbor. Studies with John Stephenson and Jacqueline Rice. Befriends Tom Phardel, a fellow student, as well as Janna Longacre and Karen Gunderman.

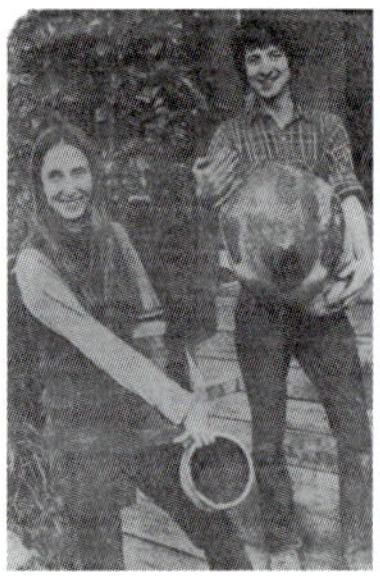

Kurt and Christy Weiser getting ready for their exhibition at the White Bird Gallery at Cannon Beach, 1974. Newspaper clipping, courtesy of the artist.

1976
Hired as graduate teaching assistant, focuses on raku vessels. Completes studies at University of Michigan, graduating with a M.F.A. Stays in Ann Arbor and is invited by Robert Piepenburg to teach a workshop at the Ann Arbor Potters Guild. Has first one-person exhibition at Artforms Gallery in Detroit, showing raku vessels inspired by pre-Columbian ceramics.

Encouraged by Ken Ferguson to apply for the position of resident director at the prestigious Archie Bray Foundation in Helena, Montana. Kurt is hired at the age of twenty-six, succeeding David Cornell. Moves to Montana in September to work along side Cornell.

1977
Assumes directorship of Bray and begins revamping the clay business operation in order to generate necessary funds to run programs. In the ensuing years, invites a number of KCAI alumni to be residents including John and Andrea Gill, Chris Gustin, David Nelson, Richard Notkin, Allan Winkler, and Arnold Zimmerman, among others. Akio Takamori comes as a resident after completing his M.F.A. from Alfred University.

1978
Continues to produce a wide range of functional pottery and raku vessels in order to support the foundation's fledgling operations. Akio Takamori, whom Kurt first befriended by their association with the Kansas City Art Institute, comes to work at the Bray, the first of four residencies. Hires Chip Clawson, the former ceramics technician at the University of Michigan to be Clay Business Manager.

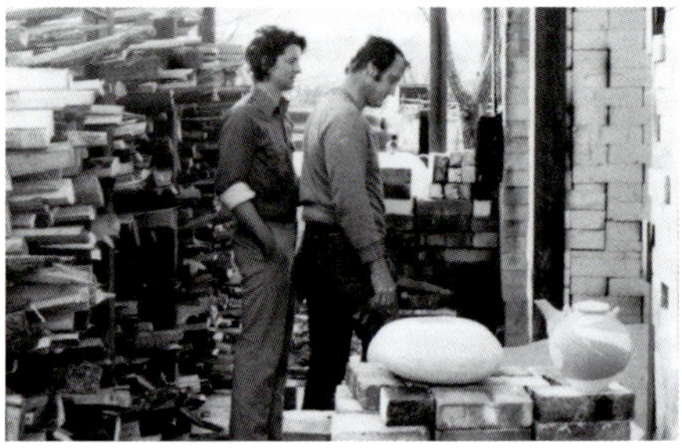

Kurt Weiser and David Shaner firing the wood kiln at the Archie Bray Foundation, 1983.

1979
Son Garth is born July 13th. Has second solo exhibition at the White Bird Gallery in Canon Beach, Oregon. Exhibits a range of stoneware and raku-fired vessels.

1981
Garth Clark, a noted ceramic historian invites Kurt to participate in a benefit auction for the Ceramic Arts Foundation in New York City; Kurt, who never heard of Clark, is reassured by Ken Ferguson and submits a plate. Wins top cash award in the Marietta College Crafts National for his raku *Toucan Jar*, providing the opportunity for a one-person show at Joanne Rapp's Hand and the Spirit Gallery in Scottsdale, Arizona.

1982
Garth Clark comes to lecture at the Archie Bray Foundation in the summer and sees a broader range of Kurt's work. Clark invites him to show in October in his recently opened Los Angeles gallery, beginning a long-time relationship that continues today. Kurt exhibits his slip-cast porcelain vases with black matte glaze and a series of salt-fired earthenware platters.

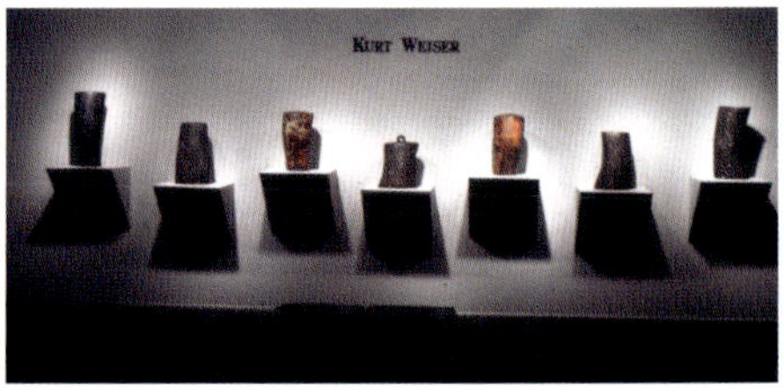

Installation of the artist's first solo exhibition at Garth Clark Gallery, Los Angeles, 1982

1983
Solo exhibition at Surroundings Gallery in New York.

1984
Negotiates the purchase of the old Bray brickyards and adjacent 13 acres, with the assistance with William Porter, the foundation's board president. This allows for the residency program to expand and provide for site specific projects to be constructed.

Invited to participate in the group exhibition *Architectural Ceramics: Eight Concepts*, which proves to be challenging. The artist glazes large square kiln shelves for his wall mounted project.

1985
Tomoko and Kazz Nakamoto come for a one-year residency. Acquaintances of Akio Takamori, Tomoko fabricates contemporary kimonos made from antique Japanese fabrics while Kazz makes a series of large-scale, low-relief tiled portraits.

1986
Receives Visual Artists Fellowship Grant from the Montana Arts Council, allowing him to travel to Japan for the first time; Bray resident Yumi Kiyose accompanies him. Visits the International Ceramics show in Mino and Goro Suzuki's studio as well as many local potteries. The experience deepens his appreciation for both historical and contemporary Asian ceramics.

1987
In May, Kurt invites Japanese ceramist Kichizaemon Raku , whom he met the previous year, to lecture on his family's history with the raku process at the Bray.

1988
In December, relocates to Tempe, Arizona, to start teaching at Arizona State University. The desert environment seems alien after his Bray tenure in rural Montana.

1989
Begins teaching in spring semester, replacing retiring professor emeritus Donald Schaumburg. Fellow colleagues include Jeanne Fronske Otis and Randall Schmidt. Receives his first National Endowment for the Arts Visual Artists Fellowship Grant, using funding to build a studio adjacent to his home. Starts to make work similar to his Bray production but becomes dissatisfied.

Inspired by the condensed information found on cereal boxes, commences a radical departure of new body of work comprised of black-stained cast porcelain teapots with graffito drawings. Former Bray resident Louis Katz travels to Thailand on a Fulbright grant, Kurt joins him later that summer. Weiser is guided by Suwanee Natewong, a Thai potter who was a resident artist at the Bray in 1987. Is overwhelmed by the flora and fauna.

1990
Returns and continues making slip-cast porcelain teapots with sgraffito designs inspired by his recent travels. Has first solo exhibition at the Garth Clark Gallery in New York, winning critical acclaim.

1991
The National Council of Education in the Ceramic Arts (NCECA) meets in Tempe, Arizona. Kurt invites several Thai potters to demonstrate.

1992
Receives 2nd National Endowment for the Arts Visual Artists Fellowship Grant. Exhibits his first series of colored china-painted vessels at Garth Clark Gallery in Los Angeles in a two-person show with Chris Staley.

1993–95
The artist produces a large body of china-painted vessels, which garners attention from collectors and curators worldwide. Participates in five solo exhibitions and over twenty group national and international shows.

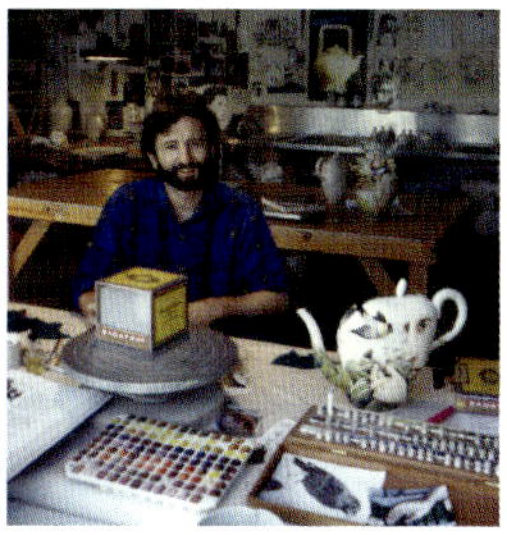

The artist in his Tempe, Arizona, studio, 1993.

1996
Receives sabbatical leave from teaching and spends time traveling to Japan.

1998
Receives an Artists Fellowship Grant from the Asian Arts Council. Returns to Japan and visits major pottery sites including Bizen, Imari, Mashiko, Tokaname. Returns and has first exhibition with the Frank Lloyd Gallery, Santa Monica, California, exhibiting china painted teapots and jars.

Travel and international residencies has informed the artist's work for the last two decades. Imari, Japan, 1998.

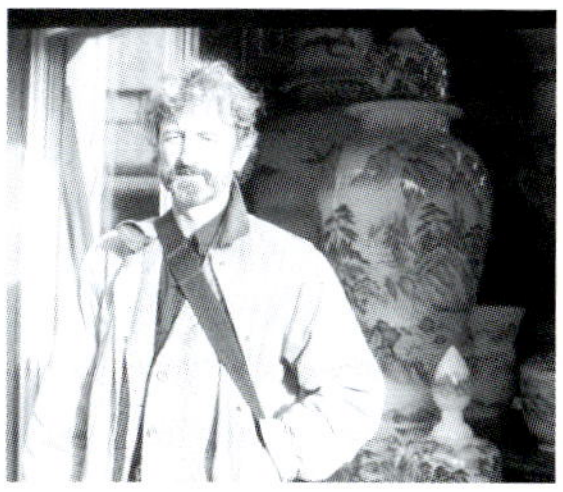

1999
Awarded a Regents Professorship at Arizona State University, the first professor in the School of Art to receive this honor. Receives an Artists Fellowship Grant from the Arizona Commission on the Arts. Son Garth enrolls at Cooper Union to study painting.

Has first solo museum exhibition *Kurt Weiser: Working His Way Around China* at the Montgomery Museum of Fine Arts, Montgomery, Alabama.

2002
Participates in the exhibition Contemporary American Ceramics at the National Museum of Modern Art in Kyoto, Japan.

2003
Receives the prestigious Aileen Osborn Webb Award, inducted in the College of Fellows by the American Craft Council. Garth Clark introduces Kurt at the awards ceremony held in conjunction with SOFA Chicago. Receives the Path of Inspiration Distinguished Alumni Award from Interlochen Academy in New York City.

2004
Begins a six-month sabbatical, working at the Royal Copenhagen Factory. Travels to Spain and Portugal visiting museums; especially impressed with the collections at the Prado, Madrid (awed by masterpieces of Velázquez, Goya, and El Greco); Thyssen-Bornemisza (primitive Flemish paintings). Goes to Barcelona and visits the Mirò Museum and Barcelona Ceramics Museum. Travels to Granada to see the Alhambra, also visiting Seville where he is overwhelmed by the architecture.

2005
Opens his groundbreaking exhibition, *Globes*, at the Garth Clark Gallery in September, signaling a new departure from his previous vessel oriented work. Son Garth graduates with an M.F.A. in painting from Columbia University, New York.

2006
Provides leadership for the planning and remodeling of new graduate ceramics studios at Arizona State University. Jeanne Fronske Otis and Randall Schmidt retire from teaching; Susan Beiner and Sam Chung are hired as new ceramics faculty.

2007
Receives the Meloy-Stevenson Award of Distinction from the Archie Bray Foundation.

Biography and Selected Exhibition History

Born

Lansing, Michigan, **1950**

Education

1967–1969

Attends Interlochen Arts Academy, Interlochen, Michigan

1969–1972

Studied one semester at the Detroit Society of Arts & Crafts, Detroit, Michigan. B.F.A., Kansas City Art Institute, Kansas City, Missouri

1974–1976

M.F.A., University of Michigan, Ann Arbor

Studios

1972–1974

Maintained a studio in Portland, Oregon

1977–1988

Maintained a studio in Helena, Montana, Archie Bray Foundation

1988 to present

Maintains a studio in Tempe, Arizona

Awards

1986

Visual Artist Fellowship, Montana Arts Council

1989

Visual Artist Fellowship, National Endowment for the Arts

1992

Visual Artist Fellowship, National Endowment for the Arts

1998

Artist Fellowship Grant, Asian Cultural Council; Research and Creativity Award, Arizona State University

1999

Regents Professorship, Arizona State University; Artist Fellowship Grant, Arizona Commission on the Arts

2003

Path of Inspiration Award, Interlochen Arts Academy; Aileen Osborn Webb Award, College of Fellows, American Craft Council

2007

Meloy-Stevenson Award of Distinction, Archie Bray Foundation

Selected One-Person Exhibits*

*catalogue published

1976

Kurt Weiser, Artforms Gallery, Detroit, Michigan.

1981

White Bird Gallery, Cannon Beach, Oregon

1982

Kurt Weiser, Surroundings Gallery, New York.

Kurt Weiser: New Works, Garth Clark Gallery, Los Angeles, California.

The Craftsmens Gallery, Scarsdale, New York.

1983

Kurt Weiser, The Hand and the Spirit Gallery, Scottsdale, Arizona.

Kurt Weiser: New Works, Craftsman Gallery, Omaha, Nebraska.

Masterful Strokes in Clay, Brentwood Gallery, St. Louis, Missouri.

1984

Kurt Weiser, Anderson Ranch Art Center, Aspen, Colorado.

Kurt Weiser, Lawrence Gallery, Portland, Oregon.

Kurt Weiser, Yellowstone Art Center, Billings, Montana.

Cohen Gallery, Denver, Colorado.

1985

Kurt Weiser, Paris Gibson Square Art Center, Great Falls, Montana.

White Bird Gallery, Cannon Beach, Oregon.

1986

Kurt Weiser, Lawrence Gallery, Portland, Oregon.

Kurt Weiser, Ceramics, Salem Art Association, Salem, Oregon.

1987

Kurt Weiser, Hui Noeau Center for Visual Arts, Maui, Hawaii.

1990

Kurt Weiser, Garth Clark Gallery, New York.

1992

Kurt Weiser: Recent Works, Garth Clark Gallery, Los Angeles, California.

1993

Kurt Weiser, Garth Clark Gallery, New York.

Kurt Weiser: China Painted Porcelain, The Hand and the Spirit Gallery, Scottsdale, Arizona.

1994

Kurt Weiser, Garth Clark Gallery, Los Angeles, California.

1995

Kurt Weiser, Garth Clark Gallery, New York,.

1996

Garth Clark Gallery, New York.

The Hand and the Spirit Gallery, Scottsdale, Arizona.

1997

Kurt Weiser, Garth Clark Gallery. New York.

1998

Kurt Weiser, Frank Lloyd Gallery, Santa Monica, California.

1999

Kurt Weiser: Working His Way Around China, Montgomery Museum of Fine Arts, Montgomery, Alabama.*

2000

Kurt Weiser, Garth Clark Gallery, New York.*

2001

Frank Lloyd Gallery, Santa Monica, California.

2002

Kurt Weiser, Garth Clark Gallery, New York.

2005

Kurt Weiser: Globes Garth Clark Gallery, New York.

55 Cups, Holter Museum of Art, Helena, Montana

2007

*Eden Revisited: The Ceramic Art of Kurt Weiser.** National Tour.

Selected Group Ehxibitions

*catalogue published

1977

Functional Ceramics, College of Wooster Art Museum, Wooster, Ohio.*

1979

Clay, Fiber, Metal, Southeastern Massachusetts University, North Dartmouth.* Award.

Functional Forms '79, Fairbanks Gallery, Oregon State University, Corvallis.*

1980

Clay Communications, Louisville School of Art, Louisville, Kentucky. *

1981

12 from Kansas City + 1: Ken Ferguson, Surroundings Gallery, New York, New York.*

Fire and Ashes, Pittsburgh Center for the Arts, Pittsburgh, Pennsylvania. Regional Tour.

Marietta College Crafts National '81, Marietta College, Ohio.* First Prize Judges Award of Craftsmanship. Regional Tour.

Northen Rockies Clay, Hockaday Center for the Arts, Kalispell, Montana.

Teapots, U.S.A., Southern Illinois University, Edwardsville.

The Art of Tea, Detroit Gallery of Contemporary Crafts, Detroit, Michigan.

Lodestone Gallery, Boulder, Colorado.

1982

All Clay: Ferguson, Ford, Weiser, Cheney Cowles Memorial Museum, Spokane, Washington.

American Clay II: Elements of Direction, Meredith Contemporary Art, Baltimore, Maryland.*

Animal Forms, Meyer Breier Weiss Gallery, San Francisco, California.

Crafts of the Northwest, Northwest Craft Center, Seattle, Washington.

Ninth Anniversary Exhibit: The Art of Clay, The Craftsman Gallery, Scarsdale, New York.

Raku, Lawrence Gallery, Gleneden Beach, Oregon.

Somers Invitational, '82 Ceramic Art, Somerstown Gallery, Somers, New York.

Transfigured Surfaces, Garth Clark Selects, Brentwood Gallery, St. Louis, Missouri.

Variations in Clay, GMB Gallery, Birmingham, Michigan.

1983

A Personal View: Selections from the Joan Mannheimer Collection, University of Missouri, Kansas City.*

Clay Invitational, University Gallery, University of Wisconsin, River Falls.

Clay Spectrum, The Craftsman Gallery, Scarsdale, New York.

Ceramic Artists: Distinguished Alumni of the Kansas City Art Institute Charlotte Crosby Kemper Gallery, Kansas City Art Institute, Kansas City, Missouri.*

Frank Boyden, Kurt Weiser: Ceramics, Northwest Crafts Center, Seattle, Washington.

Horsley/Weiser, The American Hand, Washington, D.C.

Impressions, Patterns, Arrowmont Gallery, Gatlinburg, Tennessee.

Masters in Clay, American Art Incoporated, Atlanta, Georgia.

Ten Year Anniversary Show, Sun Valley Center for the Arts and Humanities, Ketchum, Idaho.

Three Statements in Clay, Swan Gallery, Philadelphia, Pennsylvania.

1984

10th Biennial National Invitational Crafts Exhibition, University Museums Center for Visual Arts, Illinois State University, Normal.*

American Ceramics, Elaine Potter Gallery, San Francisco, California.

Architectural Ceramics: 8 Concepts, Washington University Gallery of Art, St. Louis, Missouri.* Regional Tour.

Clay – 84, Traver Sutton Gallery, Seattle, Washington.

Multiplicity in Clay, Metal, Fiber, Skidmore College Art Center, Saratoga Springs, New York.*

Platters and Related Forms, Southern Illinois University, Edwardsville.*

1985

American Clay: 1960–1985, Panhandle-Plains Museum, Canyon, Texas.*

Ceramic Implications For Architecture, The Hand and the Spirit Gallery, Scottsdale, Arizona.

Kansas City Art Institute's First Century, Nelson-Atkins Museum of Art, Kansas City, Missouri.

NCECA Members Juried Exhibition, Craft Alliance Center for the Visual Arts, St. Louis, Missouri.

Sanford M. Besser Collection of Contemporary Teapots, Arkansas Arts Center Decorative Arts Museum, Little Rock, Arkansas.*

Three Ceramic Artists, Contemporary Crafts Gallery, Portland, Oregon.

1986

American Potters Today, Victoria and Albert Museum, London, England.*

Clay AZ Art, Gummesons Art Gallery, Stockholm, Sweden. International Tour.

Teapots From the Archie Bray Foundation, Paxton Gallery, University of Montana, Missoula.

1987

A Gathering of the Crafts, Gallery of the Honolulu Academy of Arts, Honolulu, Hawaii.

Celebrating Craft, Detroit Gallery of Contemporary Crafts, Detroit, Michigan.

Ceramic Sculpture Exhibition, New Mexico State University, Las Cruces.

Clay Today: Contemporary Ceramists and Their Work, Los Angeles County Museum of Art, Los Angelas, California.*

Montana Visual Arts Fellowship Recipients, Hockaday Arts Center, Kalispell, Montana

1988

A Fine Place to Work: The Legacy of the Archie Bray Foundation, Arkansas Arts Center Decorative Arts Museum, Little Rock.

Archie Bray International: Summer Residents, Holter Museum of Art, Helena, Montana.

Northwest Ceramics Today, Cheney Cowles Memorial Museum, Spokane, Washington.* Regional Tour.

Northwest Clay, Coos Art Museum, Coos Bay, Oregon. Regional Tour.

The Arts: Montana's Cultural Ambassador to Japan, The Ginza Gallery, Tokyo, Japan.

1989

10 Montana Artists: Art That Crosses all Boundaries, Artworks Gallery, Seattle, Washington.

Clay Arizona Art Conference Exhibit, Northern Arizona University Art Gallery, Flagstaff.

The Eccentric Teapot, Garth Clark Gallery, New York.*

Summer of 88—Archie Bray Foundation, Central Park Gallery, Kansas City, Missouri.

White Heat: A Collection of Porcelain Ceramics, Gallery of the Department of Art and Art History, University of Nebraska, Lincoln.

1990

A Tea Party, Ferrin Gallery, Northampton, Massachussetts.

Black and White: Contemporary Ceramics, The Hand and the Spirit Gallery, Scottsdale, Arizona.

Clay, Color, Content: 28th Ceramic National, Everson Museum of Art, Syracuse, New York.* National Tour.

Finding 90: A Recognition of Ceramic Excellence, The Society of Arts and Crafts, Boston, Massachussetts.

High Tea, Moira James Gallery, Green Valley, Nevada.

NCECA Juried Members Exhibition, Taft Museum, Cincinatti, Ohio.* National Tour.

New Art Forms, Chicago International, Navy Pier, Chicago, Illinois.

Recent Fires: Contemporary American Ceramics. Utah Museum of Fine Arts, University of Utah, Salt Lake City.* National Tour.

Selections from the Joyce and Jay Cooper Collection of Contemporary Ceramics, Arizona State University Art Museum, Tempe, Arizona.*

1991

An Exhibition Celebrating Ceramists from the Gallery's First Twenty Years, The Hand and the Spirit, Scottsdale, Arizona.

Archie Bray Foundation Benefit Exhibition, Helen Drutt Gallery, New York, New York.

Clay Coast to Coast, Trisolini Gallery, Ohio University, Athens.

Ceramics '91, The Farrell Collection, Washington D.C.

Fired By Imagination: Clay Today, Mint Museum of Art, Charlotte, North Carolina.*

Minor Works-Major Artists, South Mountain Center for the Arts, Phoenix, Arizona.

NCECA '91 ASU School of Art Ceramic Faculty Show, Memorial Union Fine Arts Gallery, Tempe, Arizona.

New Art Forms, Chicago Exposition, Navy Pier, Chicago, Illinois.*

Teapots: Contemporary Views, Pro Art Gallery, St. Louis, Missouri. National Tour.

The Tea Party, American Craft Museum. New York.

Works in Clay 7, Witchita Falls Museum, Witchita Falls, Texas.

1992

1992 International Invitational Exhibition of Ceramic Art, The National Museum of History, Taipei, Taiwan.

American Teapots, Shigaraki Museum, Shigaraki, Japan. Regional Tour.

Art of the Figure, Twist Gallery, Portland, Oregon.

Artistic Generations: Jacqueline Rice and Her Students, Works Gallery, Philadelphia, Pennsylvania.

Black and White II, The Hand and the Spirit, Scottsdale, Arizona.

Contemporary Views, Pro Art Gallery, St. Louis, Missouri.

David Regan, Kurt Weiser, Garth Clark Gallery, Kansas City, Missouri.

Rituals of Tea, Garth Clark Gallery, Los Angeles, California.

Sculptural Objects Functional Art, Chicago Navy Pier, Chicago, Illinois.*

Teapot Invitational, Dorothy Weiss Gallery, San Francisco, California.

1993

Built, Thrown and Touched, Media Gallery, Kansas City, Missouri. National Tour.

Contemporary American Ceramics, Duncan Gallery of Art, DeLand, Florida.

Every Object Tells a Story: Narrative Craft, Arizona State University Art Museum, Tempe, Arizona.*

Exhibition of Works Boran at the Park, Institute of Ceramic Studies, Shigaraki, Japan.

Montana Choice: Montana Arts Council Fellowship Artists 1983–1993, Hockaday Center, Kalispell, Montana.

Rituals of Tea, Garth Clark Gallery, Los Angeles, California.

Sculptural Objects Functional Art, Chicago Navy Pier, Chicago, Illinois.*

The Legacy of the Archie Bray Foundation, 4 Decades of Tradition and Innovation in American Ceramic Art, Bellevue Art Museum, Bellevue, Washington and the Archie Bray Foundation, Helena, Montana.*

The Tea Party, National Museum of Ceramic Art, Baltimore, Maryland.

The Year of the American Craft, Malveety Gallery, Gleneden Beach, Oregon.

1994

Ancient China Modern Clay, Phoenix Art Museum, Phoenix, Arizona.*

Sculptural Objects Functional Art, Chicago Nacy Pier, Chicago, Illinois.*

The Archie Bray Foundation: A Ceramic Tradition, Cultural Center for Cascade County, Great Falls, Montana.

The Archie Bray Foundation: Selected Clay Works, The Hand and the Spirit, Scottsdale, Arizona.

The Collector's Eye, The Koffler Gallery, North York, Ontario, Canada.*

1995

Distinguished Fires – A Survey of Montana Ceramics, Yellowstone Art Center, Billings, Montana.

Keepers of the Flame: Ken Ferguson's Circle, Kemper Museum of Contemporary Art and Design, Kansas City, Missouri.

Sculptural Objects Functional Art, Chicago Navy Pier, Chicago, Illinois.*

The Nude in Clay, Perimeter Gallery, Chicago, Illinois.* Regional Tour.

1996

96–97 ASU Art Faculty Exhibition, Arizona State University Art Museum, Tempe, Arizona.

Annual Teapot Exhibition, Craft Alliance Gallery, St. Louis, Missouri.

Sculptural Objects Functional Art, Chicago Navy Pier, Chicago, Illinois.*

Strong Museum, Rochester, New York.

1997

Food Glorious Food: Artists and Eating, Charles A. Wustum Museum of Fine Arts, Racine, Wisconsin.*

Sculptural Objects Functional Art, Chicago Navy Pier, Chicago, Illinois.*

Tantalizing Teapots, Craft and Folk Art Museum, Los Angeles, California.

Wondrous Teapots, Craft Alliance Gallery, St. Louis, Missouri.

1998

American Craft, Renwick Gallery, Smithsonian American Art Museum, Washington, D.C.

Sculptural Objects Functional Art, Chicago Navy Pier, Chicago, Illinois.*

Splendor of Porcelain, Los Angeles County Art Museum, LosAngeles, California.

1999

Anne & Sam Davis Collection, Arizona State University Art Museum, Tempe, Arizona.

Contained Narratives, Garth Clark Gallery, New York.

Millenial Flow, JBK Gallery, Amsterdam, Holland.

Nine Decades, Chester Springs Studio Center for Visual Art, Chester Springs, Pennsylvania.

Precious, Ceramic Millenium Arts Festival, Amsterdam, New York.

Sculptural Objects Functional Art, Chicago Navy Pier, Chicago, Illinois.*

Teapot Invitational, Dorothy Weiss Gallery, San Francisco, California.

Toronto Collects, The George Gardiner Museum of Art, Toronto, Canada.

Works on Paper, John Elder Gallery, New York.

2000

12th Yeoju Ceramic Expo, Yeoju, Korea.*

Ceramics National 2000, Everson Museum of Art, Syracuse, New York.* National Tour.

Clay at the Crossroads, Winnipeg Art Gallery, Winnipeg, Manitoba, Canada.

Color and Fire: Defining Moments in Studio Ceramics, 1950 –2000, Los Angeles County Art Museum, Los Angeles, California.* National Tour.

Defining Crafts: Collection for the New Millennium. American Craft Museum, New York, *.

Gardens of Pleasure, John Michael Kohler Arts Center, Sheboygan, Wisconsin.

Portrait Pots, Frank Lloyd Gallery, Santa Monica, California.

2001

1st World Ceramic Biennale 2001 Korea International Competition, Ichon, Korea.* Regional tour.

2001: Clay Odyssey: Celebrating 50 Years of the Archie Bray Foundation, Helena, Montana.

A Ceramic Continuum: 50 Years of the Bray Influence, Holter Museum of Art, Helena, Montana.* National Tour.

Katherine K. Herberger College of Fine Arts School of Art Faculty Exhibition, Arizona State University Art Museum, Tempe, Arizona

Birds, Beasts and Little Fishes, Santa Fe Clay, Santa Fe, New Mexico.

USA Clay, Smithsonian American Art Museum, Renwick Gallery, Smithsonian American Art Museum, Washington, D.C.

Wonder and Discovery—Contemporary Ceramic Arts of America, Museum of Contemporary Art, Shigaraki, Japan.

2002

Contemporary American Ceramics, 1950–1990, Kyoto Museum of Modern Art, Kyoto, Japan.* Regional Tour.

International Ceramics Exhibition, Yingge Ceramics Museum, Yingge, Taiwan.*

Small is Beautiful, Frank Lloyd Gallery, Santa Monica, California.

Spinning Tales: Narrative Ceramics, Sybaris Gallery, Royal Oak, Michigan.

Teapots and Opium, Garth Clark Gallery, New York.

Texas National 2002, Stephen F. Austin State University, Nacogdoches, Texas.

The Artful Teapot: 20th Century Expressions from the Kamm Collection, Curated by Garth Clark, circulated by Exhibitions International.* Interational Tour.

2003

21st Century Ceramics of the U.S. and Canada, Columbus College of Art and Design, Columbus, Ohio.*

2003 Intermountain Invitational, Nora Eccles Harrison Musuem of Art, Logan, Utah.

Great Pots: Contemporary Ceramics from Function to Fantasy, Newark Museum, Newark, New Jersey.*

Life Taking Shape: Ceramics from the Aaron Milrad Collection, Gardiner Museum of Ceramic Art, Toronto, Canada.

Sculptural Objects Functional Art, Chicago Navy Pier, Chicago, Illinois.*

Shared Passion: Sara and David Lieberman Collection of Contemporary Ceramics and Craft, Arizona State University Art Museum Ceramics Research Center, Tempe Arizona.*

Tea Anyone? The Donna Moog Teapot Collection, Racine Art Museum, Racine, Wisconsin.*

2004

Standing Room Only 2004 Scripps 60th Ceramic Annual, Scripps College, Claremont, California.*

Bare Clay, Garth Clark Gallery, New York.

From the Garden, Northern Clay Center, Minneapolis, Minnesota*

Sculptural Objects Functional Art, Chicago Navy Pier, Chicago, Illinois.*

Steamin' Teapots, Gallery Materia, Scottsdale, Arizona.

The Animal Instinct, Baltimore Clayworks, Baltimore, Maryland.

The Signature Vessel, Garth Clark Gallery, New York.

Works from the Permanent Collection, Mesa Contemporary Arts, Mesa, Arizona.

American Masters in Clay, Westmont Reynolds Gallery, Santa Barbara, California.

2005

Animated Earth: The Many Permutations of Clay, Craft Alliance, St. Louis, Missouri.

30 x 5 Annual Ceramics Invitational, Akar Design, Iowa City, Iowa.`

American Masters, Santa Fe Clay, Santa Fe, New Mexico.

Connoisseurs of Clay: Collecting Contemporary Ceramics, Museum of Fine Arts, St. Petersburg, Florida.

Innovation and Change: Great Ceramics From the Permanent Collection, ASU Art Museum Ceramics Research Center, Tempe, Arizona.

Past, Present and Future: Arizona Ceramic Artists, St. Louis Community College, St. Louis, Missouri.

Terra Nova: Sculpture and Vessels in Clay, Museum of Arts and Design, New York.

2006

Great Pots: The Vessel as Art, 1900–2000. Organized by the Newark Art Museum UBS Art Gallery, 1285 Avenue of the Americas, New York City.

Selected Works exhibited in Luce Foundation Center for American Art, Smithsonian American Art Museum, Washington, D.C.

Two Media / Two Expressions: An Exhibition of Contemporary Clay and Glass, Springfield Museum of Art, Springfield, Ohio.

Legacy and Innovation in Contemporary Clay, Dubuque Museum of Art, Dubuque, Iowa.

Poetic Expressions of Mortality. Figurative Ceramics from the Porter Price Collection, Mobile Museum of Art, Mobile, Alabama.

New Works 2006: Celebrating Recent Gifts to RAM, Racine Art Museum, Racine, Wisconsin

Our Cup Runneth Over, The Society of Arts and Crafts, Boston, Masachusetts.

Tell Me a Story, Santa Fe Clay, Santa Fe, New Mexico.

30 x 5, Akar Gallery, Iowa City, Iowa.

The Jeanne and Bill Porter Collection of Ceramic Art, Holter Museum of Art, Helena, Montana.

The Yixing Effect: Echoes of the Chinese Scholar, Holter Museum of Art, Helena, Montana

Tea Time: The Art of the Tea Pot, Kalamazoo Institute of the Arts, Kalamazoo, Michigan

2007

Cityscape/Landscape, The Clay Studio, Philadelphia, Pennsylvania.

The Diane and Sandy Besser Collection, Fine Arts Museums, deYoung Museum, San Francisco, California

Selected Public and Corporate Collections

Archie Bray Foundation for the Ceramic Arts, Helena, Montana

Arizona State University Art Museum, Ceramics Research Center, Tempe, Arizona

Carnegie Institute Art Museum, Pittsburgh, Pennsylvania

Cox Communication Corporation, Atlanta, Georgia

Crocker Art Museum, Sacramento, California

The Gardiner Museum, Toronto, Canada

Hallmark Art Collection, Kansas City, Missouri

Helsinki Museum of Applied Art, Helsinki, Finland

Kansas City Art Institute, Kansas City, Missouri

Los Angeles County Museum of Art, Los Angeles, California

Mint Museum of Craft + Design, Charlotte, North Carolina

Museum of Arts & Design, New York, New York

Museum of Contemporary Ceramics, Shigaraki, Japan

National Museum of American Art, Smithsonian Institution, Washington, D.C.

National Museum of History, Republic of China, Taipei, Taiwan

The Newark Museum, Newark, New Jersey

Nora Eccles Harrison Museum of Art, Utah State University, Logan, Utah

Racine Art Museum, Racine, Wisconsin

Rhode Island School of Design, Museum of Art, Providence, Rhode Island

SAFECO Corporation, Seattle, Washington

Schein-Joseph International Museum of Ceramic Art, Alfred, New York

Victoria and Albert Museum, London, England

Washington University Art Museum, St. Louis, Missouri

Winnipeg Art Museum, Manitoba, Canada

Yellowstone Art Museum, Billings, Montana

Selected Bibliography

Books and Catalogues

Adamson, Glenn, and Davira S. Taragin. *Tea, Anyone? The Donna Moog Teapot Collection*. Racine, Wisconsin: Racine Art Museum, 2003, p. 85.

Aichi-ken Toji Shiryokan. *Amerika gendai togei no keifu, 1950–1990: jiyu no kuni no no obuje to utsuwa (Contemporary American Ceramics, 1950–1990: A Survey of American Objects and Vessels)*. Tokyo, Japan: Nihon Keizai Shinbusha, 2002, pp. 127, 170–171.

Ashton, Dore, et al. *Multiplicity in Clay Metal Fiber: Skidmore College Craft Invitational 1984, July 7 – September 23*. Saratoga Springs, New York: Skidmore College Art Center, 1984, pp. 7, 37, 86, 105.

Bloemink, Barbara J., et al. *Keepers of the Flame: Ken Ferguson's Circle, Mar. 17 – May 21, 1995*. Kansas City, Missouri: Kemper Museum of Contemporary Art and Design of Kansas City Art Institute, 1995, pp. 30, 40, 51, 63–64, 73.

Boyd, Karen Johnson, et al. *The Nude in Clay*. Chicago, Illinois: Perimeter Gallery, 1995, pp. 15, 31, 33.

Brown, Claudia, and Robert Mowry. *Ancient China Modern Clay: Chinese Influences of 5 Ceramic Artists*. Phoenix, Arizona: Phoenix Art Museum, 1994, pp. 21–24, 29–31.

Brown, Claudia, *Kurt Weiser: Working His Way Around China*. Montgomery, Alabama: Montgomery Museum of Fine Arts, 1999.

Clark, Garth. *The Artful Teapot*. New York: Watson-Guptill Publications, 2001, pp. 156–157, 159, 226, 248.

———. New York: Garth Clark Gallery, 2000.

Clark, Garth, and Watson, Oliver. *American Potters Today*. London: Victoria & Albert Museum, 1986, pp. 53, 84.

Cohen, Cathy. *Anderson Ranch Arts Center Faculty Exhibit '83*. Denver, Colorado: Cohen Gallery, 1983, p. 3.

Coleman, Tom. *With a Little Help from My Friends*. Utah: Tom Coleman, 2000, pp. 26–27.

Del Vecchio, Mark. *Postmodern Ceramics*. New York: Thames and Hudson, 2001, pp.126–127, 218–219.

Dietz, Ulysses Grant. *Great Pots, Contemporary Ceramics from Function to Fantasy*. Madison, Wisconsin: Guild Publishing, 2003, pp 142–143, 151–152.

Ferguson, Ken. *Ceramic Artists: Distinguished Alumni of the Kansas City Art Institute: October 8 – November 6, 1983, Charlotte Crosby Kemper Gallery*. Kansas City, Missouri: The Kansas City Art Institute, 1983, pp. 13–14, 30.

Ferrin, Leslie. *Teapots Transformed, Exploration of an Object*. Madison, Wisconsin: Guild Publishing, 2000, p. 92.

Garten, Cliff. *Architectural Ceramics: Eight Concepts*. St. Louis, Missouri: Washington University Gallery of Art, 1985, pp. 24–25.

Gibson, John. *Contemporary Pottery Decoration*. Radnor, Pennsylvania: Chilton Book Company, 1987, p. 109.

Gunter, Veronika Alice. *500 Figures in Clay: Ceramic Artists Celebrate the Human Form*. New York, New York: Lark Books, 2004, p. 396.–397.

Hardin, Jennifer, and Peter Held. *Connoisseurs of Clay: Collecting Contemporary Ceramics*. St. Petersburg, Florida: Museum Of Fine Arts, 2005, pp. 2, 8,12, 18.

Harrington, La Mar. *Ceramics in the Pacific Northwest: A History*. Seattle: University of Washington Press, 1978, p. 30.

Held, Peter, et al. *A Ceramic Continuum: Fifty Years of the Archie Bray Influence*. Seattle: University of Washington Press and Helena, Montana: Holter Museum of Art, 2001, pp. 41–62, 63–76, 96, 104.

Held, Peter. *Between Clouds of Memory: Akio Takamori, A Mid-Career Survey*. Tempe: Arizona State University Art Museum, 2005, pp. 14, 16, 28, 35, 127–128.

Held, Peter, and Susan Peterson. *Shared Passion: Sara and David Lieberman Collection of Contemporary Ceramics and Craft*, Tempe: Arizona State University Art Museum, 2003, pp. 19, 24, 26.

Herman, Lloyd. *The Collector's Eye: American and British Ceramics from the Collection of Aaron Milrad*. North York, Ontario: Koffler Gallery, 1994, p. 55.

———. *Northwest Ceramics Today*. Boise, Idaho: Boise State University, 1987, p. 47.

Hopper, Robin. *The Ceramic Spectrum: A Simplified Approach to Glaze and Color Development, 2nd edition*. Iola, Wisconsin: Krause Publications, 2001, p. 141.

———. *Contemporary Studio Ceramics, Collection of Winnipeg Art Gallery*. Winnipeg, Canada: Winnipeg Art Gallery, 1997, p. 222.

Hunt, Bill. *21st Century Ceramics in the United States and Canada*. Westerville, Ohio: American Ceramics Society, 2003, p. 222.

Kirsch, Elisabeth, et al. *A Personal View: Selections from the Joan Mannheimer Collection*. Kansas City, Missouri: University of Missouri—Kansas City, 1983, pp. 35–36.

Lauria, Jo. *Color and Fire—Defining Moments in Studio Ceramics 1950–2000*. New York: Rizzoli International Publications, 2000, p 69.

Leach, Mark. *Fired By Imagination: Clay Today*. Charlotte, North Carolina: Mint Museum of Art, 1991, pp. 2–4, 6.

Levin, Elaine. *The History of American Ceramics 1607 to the Present, from Pipkins and Bean Pots to Contemporary Forms*. New York: Harry N. Abrams, Inc., 1988, p. 323.

Lynn, Martha Drexler. *Clay Today, Contemporary Ceramists and their Work: A Catalog of the Howard and Gwen Laurie Smits Collection at the Los Angelas County Museum of Art*. Los Angeles, California: Los Angeles County Museum of Art, 1990, pp. 160–161, 222.

MacNaughton, Mary Davis, et al. *Standing Room Only, 2004 Scripps 60th Ceramic Annual*. Claremont, California: Scripps College, 2004, pp. 162, 171.

Morganthal, Deborah, and Suzanne J.E. Tourtillott. *Penland Book of Ceramics, Master Classes in Ceramic Techniques*. New York: Lark Books, 2003, p. 8.

Museum of Contemporary Ceramic Art, The Shigaraki Ceramic Cultural Park. *Catalogue of the Collection, 1992*. Shiga Prefecture: Museum of Contemporary Ceramic Art, The Shigaraki Ceramic Cultural Park, 1993, pp. 250–251.

Ostermann, Matthias. *The Ceramic Surface*. Philadelphia: University of Pennsylvania Press, 2002, pp. 156–157.

Perry, Barbara, ed. *Clay, Color, Content The 28th Annual Ceramic National*. Syracuse, New York: Everson Museum of Art, 1990, pp. 9, 82.

Peters, Lynn. *Surface Decoration for Low Fire Ceramics*. Asheville, North Carolina: Lark Books, 1999, p. 63.

Peterson, Susan. *Contemporary Ceramics*. New York: Watson-Guptill Publications, 2000, p. 175.

———. *The Craft and Art of Clay*. Woodstock, New York: Overlook Press, 2000, p. 326.

———. *Working With Clay: An Introduction.* Woodstock, New York: Overlook Press, 1998, p. 101.
Piché, Thomas, and Justin Clemens. *Everson Ceramics National 2000: The 30th Ceramic National Exhibition.* Syracuse, New York: Everson Museum of Art, 2000, pp. 35, 84, 105–106.

Piepenburg, Robert. *Raku Pottery.* New York: Macmillan Company, 1972, p.105.

Preu, Nancy, et al. *Defining Craft 1: Collecting for the New Millenium.* New York: American Craft Museum, 2000, p. 61.

Rawson, Philip. *Recent Fires: Contemporary American Ceramics.* Salt Lake City: Utah Museum of Fine Arts, 1990, pp. 6, 16.

Rhodes, Daniel. *Clay and Glazes for the Potter, 3rd edition.* Iola, Wisconsin: Krause Publications, 2000, pp. 168–169.

Rubin, Michael. *Architectural Ceramics: 8 Concepts.* St. Louis, Missouri: Washington University Gallery of Art, 1985, pp. 5, 9–10, 24–25.

Scott, Paul. *Painted Clay, Graphic Arts and the Ceramic Surface.* New York: Watson-Guptil Publications, 2000, pp.110–111.

Senska, Frances, and Diane Douglas. *The Legacy of the Archie Bray Foundation: Four Decades of Tradition and Innovation in American Ceramic Art.* Bellevue, Washington: Bellevue Art Museum, 1993, pp. 12–14, 27–28, 30.

Silberman, Robert. *From the Garden: Forms, Images and Ideas.* Minneapolis, Minnesota: Northern Clay Center, 2004, pp. 11–12, 18.

Surroundings Gallery. *12 From Kansas City + 1: Ken Ferguson.* New York: Surroundings Gallery, 1981, p. 15.

Williams, Gerry, et al. *The Studio Potter 20th Anniversary Collection.* Golfstown, New Hampshire: Studio Potter, 1992, p. 26.

WOCEK Organizing Committee. *Yeoju International Ceramic Workshop and World Ceramic Exhibition.* Yeoju, Korea: Yeoju University, 2001, p. 13.

Wu, Chin-Fong, et al. *The International Art Teapot Exposition: Ceci n'est pas une exposition de théières.* Yingge, Taiwan: Taipei County Yingge Ceramics Museum, 2002, pp. 106, 126.

Zakin, Richard. *Ceramics: Mastering the Craft.* Radnor, Pennsylvania: Chilton Book Company, 1990, p. 215.

Periodicals and Reviews

"2003 Aileen Osborn Webb awards." *American Craft* 63 (October/November 2003): 58–67.

"All Clay." *American Craft* (April/May 1982): 67.

"American Earthenware." *Ceramics Monthly* (May 1982): 58.

"Americans Abroad." *American Craft* (December/January 1991), 36–39.

Ballis, Kosmas. "The Challenge of Abstract Sculpture." *Ceramics Technical* 18 (2004) 8–12.

Britton, Sheilah. "Touching Mystery." *Arizona State University Research Magazine* (Summer 2002): 2–5.

Chen, Jingliang. "Master of China Painting, Kurt Weiser." *Purple Sands* 36 (September, 1996): 31–33.

Clark, Garth. "Garth Clark: Life and Works." *Studio Potter* 30 (December 2001): 3–16.

———. "Postmodern Ceramics." *Kerameiki Techni* (December 2001) 16–22.

Clowes, Jody. "Foot, Belly, Shoulder, Lip." *American Craft* 56, no.3 (June/July 1996): 52–57.

Curd, Bryan M. "Kurt Weiser: Evolving Vision." *Ceramics Art and Perception*, no. 52 (June 2003): 56–61.

Dauer, Paul F. "Color and Fire: Ceramics at LACMA." *Ceramics Art and Perception* 44 (2001): 68–73.

Degener, Patricia. "Celebrating Ken Ferguson..." *American Craft* (October/November 1995): 64–69.
"Evolution at the Bray." *Ceramics Monthly* (Summer 1985): 87–89.

Hackman, Kate. "The Power of Clay." *Kansas City Star*, December 17, 2000.

"Eclectic Collector." *New York Magazine* (September 2002): 18.

Ellison, Robert. "Philadelphia Rolls out the Clay." *American Craft* 52, no. 3 (June/July 1992) 36–45, 66.

"Everson Ceramic National 2000." *Ceramica*, no. 76 (2001) 32–35.

"Evolution at the Bray." *Ceramics Monthly* (June/July/August 1985): 87–89.

Ferrin, Leslie. "Teapots Transformed, Exploration of an Object." *Kerameiki Techni* (April 2001): 17–22.

Glueck, Grace. "Art: Using Ceramics to Adorn Architecture." *New York Times*, July 12, 1985.

Goldgate, Steven. "Working with Paperslip." *Ceramics Technical* no. 13 (2001): 78–80.

Gomez, Edward M., and Rebecca Dimling Cochran. "Top 100 Collectors." *Art and Antiques* (March 2005): 97–127.

Grainlick, Herbert. "Steinberg Architectural Ceramics Show a Good Bet to Raise Questions Aplenty." *St. Louis Globe-Democrat*, April 13, 1985.

Gyung, Choi Eun. "After Finishing Work at the Department of Ceramics of ASU—Kurt Weiser: His Life and Art." *Ceramic Art Monthly* (April 2001): 45–47.

Hackman, Kate. "The Power of Clay." *The Kansas City Star*, December 17, 2000.

Held, Peter. "The Ceramics Research Center at Arizona State University." *Ceramics Technical* 17 (2003): 102–105.

Hoffman, Donald. "A Reunion on Canvas." *Kansas City Star*, April 28, 1985.

Hunt, William. "Bone China." *Studio Potter* 26, no. 2 (June 1998): 33–35.

"In View" *Ceramic Review* (U.K.) no. 128 (March/April 1991): 9–11, 31.

Jensen, Robert. "Architectural Ceramics: Eight Concepts." *American Craft* 45 (June/July 1985): 46–51.

Kangas, Mathew. "Kurt Weiser." *American Ceramics* 2, no. 2 (Spring 1983): 70–71.

Kangas, Mathew. "Kurt Weiser." *American Ceramics* 9, no. 1 (Spring 1991): 51.

Keller, Julie. "15 Artists to Watch." *Art Business News* 31, no. 2 (February 2004): pp. 62–64, 66, 70, 72, 74–75.

Keller, Tobin. "Stan Welsh's Mixed Media Sculpture." *Ceramics*, no 37 (1999): 59–63.

"Kurt Weiser." *Phoenix Home and Gardens.* (March 2000): 163.

Landman, Roberta. "Clay Feat." *Phoenix Home and Garden* (September 2004) 60–65.

Lauria, Jo. "Dialogues in Clay: A Conversation between Tony Marsh and Kurt Weiser." *Ceramics Art and Perception*, no. 50 (December 2002): 8–13.

Leach, Mark. "Kurt Weiser." *American Ceramics* 11, no.1 (November 1993): 24–33, cover.

Lebow, Edward. "Glaze of Glory," *New Times*, March 24, 1993.

———. "Kurt Weiser, Storied Forms." *American Craft* 54 (December 1994/January 1995): 44–47.

Levin, Elaine. "Painting on China." *Ceramics Technical* 12 (2001): 14–21.

Lovelace, Joyce. "The Ubiquitous Teapot." *American Craft* 54, no. 2 (April/May 1994): 54–61.

"Marietta College Crafts National." *American Craft* (December 1981/January 1982): 6.

Matsuo, Amiko. "Reflections on Biculturalism: Japanese Ceramic Artists in the US." *Ceramics Art and Perception* 60 (2005) 31–35.

"NEA Crafts Fellowships." *Ceramics Monthly* 41, no. 2 (February 1993): 46–51.

Pyne, Lynn. "Governor's awards to recognize artists." *The Phoenix Gazette*, March 14, 1990.

Rainbolt, Jo. "Eleven Montana Potters." *Studio Potter* 8, no. 2 (1979): 44–46.

Reif, Rita. "Keeping up with the Expanding Meaning of Craft." *New York Times*, February 13, 2000.

Rubin, Michael G. "Concepts in Architectural Ceramics" *NCECA Journal* (1985): 13.

———. "New Pottery at Brentwood." *St. Louis Globe-Democrat*, January 1–2, 1983.

———. "Two Contrasting Artists Rising in Prominence." *St. Louis Globe-Democrat*, June 25–26, 1983.

Sartorius, Tara Cady. "Eden Revisited." *Arts and Activities* (February 2000): 26–28.

Schmidt, James R. "Teapots USA." *Ceramics Monthly* (November 1981): 38–39.

Senska, Frances. "Pottery in a Brickyard." *American Craft* (February/March 1982): 32–35.

Shaner, David. "The Archie Bray Foundation." *Studio Potter* 8, no. 2 (1979): 47–51

Starr, Irma. "The Renwick and Me." *Ceramics Monthly* 52, no. 4 (April 2004): 54–55.

"Steamin', A Survey of the Contemporary Teapot." *Clay Times* 10, no. 5 (September/October 2004): 29.

"Sun Valley Exhibition." *Ceramics Monthly* (February 1983): 52.

"Surreal Ceramic." *Arizona State University Research Magazine* (Fall 2000): 11.

Szabat, A.M. "'White Heat' shows porcelain in exotic variety of vessels." *Lincoln, Nebraska Journal*, October 17, 1989.

Takeuchi, Sam. "Report of International Clay Work in America: Peter Callas Workshop." *Ceramic Art and Craft Work 68* (2001): 106–109.

"Teapot." *Ceramics Monthly* 41 (February 1993): 48.

Tirrell, Norma. "The Archie Bray Legacy: Four Decades of American Ceramic Art." *Ceramics Monthly* 41, no. 10 (December 1993): 47–62.

Vivas, Antonio. "El Barro Desnudo." *Ceramica*, no. 95 (2005): 26–30.

Weiser, Kurt. "The Fire's Unpredictable Nature." *Studio Potter* 11, no. 2 (June 1983): 42.

"Wildlife." *American Craft* 56, no. 5 (October/November 1996): 31.

Williams, Gerry, et al. "Panel Discussion: In the Beginning." *Studio Potter* 30 (December 2001): 26–35.

Wilson, William. "'Tantalizing Teapots a Feel-Good Brew." *Los Angeles Times*, October 6, 1997.

Xu, Wenqin. "Multi Cultural Ceramics in 1994." *Ceramic Art* 4 (Summer 1994): 52–59.

Videos

Color and Fire—Defining Moments in Studio Ceramics 1950–2000. Directed by Elvin Whitesides, Los Angeles, California, 2000. 50 minutes.

Lenders to the Exhibition

Dr. and Mrs. George Andros

Archie Bray Foundation for the Ceramic Arts

Arizona State University Art Museum

Marvin Gerstin

Michael Gilbert & Catherine Shaner Gilbert

Roberta and Bruce Hammer

John Hoffman

Stephen & Pamela Hootkin

Barbara and Mike Jenson

Kamm Teapot Foundation

Steven Korff & Marcia Van Wagner

Dr. and Mrs. Youk Lee

Saul E. Levi

Sara & David Lieberman

Jim & Vicki Mercer

The Newark Museum

Jeanne and William Porter

Private Collection

Sandra Schwalb

Jerome Shaw

Craig Smith

Pam Smith

Renwick Gallery of the Smithsonian American Art Museum

Sprint Nextel Art Collection

Barbara and Donald Tober

Kurt and Christy Weiser

Barbara Wild

Arizona State University Art Museum Staff

Administration

Marilyn A. Zeitlin, *Director/Chief Curator*

Kathleen T. Wacker, *Business Operations Manager*

Dawne Walczak, *Accountant*

Susan Ables, *Administrative Assistant*

LaReal Eyring, *Museum Store Manager*

Curatorial

Heather Sealy Lineberry, *Senior Curator*

Peter Held, *Curator of Ceramics*

John D. Spiak, *Curator of University and Community Projects*

Jean Makin, *Print Collections Manager*

Andrea Feller, *Acting Curator of Education*

Peter Bugg, *Student Docent Coordinator*

Rudy Navarro, *Curatorial Assistant*

Anne Sullivan, *Chief Registrar*

Laura Wenzel, *Assistant Registrar*

Elisa Benavidez, *Registrarial Assistant*

Stephen Johnson, *Chief Preparator*

Fausto Fernandez, *Preparator*

Chris Miller, *Preparator*

Security

Michael Brennan, *Security Officer*

Michael Haney, *Security Officer*

Jason Ripper, *Security Officer*

Robert Bargman, *Security Officer*

Joseph Bernal, *Security Officer*

Student Assistants/Interns:

Susan E. Armstrong, *CRC & Business Office Intern*

Ann Marie Baiocchi, *Visitor Services Student Asst.*

Celeste Barajas, *Business Office Student Asst.*

Crystal Barajas, *Visitor Services Student Asst.*

Janelle Cairo, *Intern*

Peter Bugg, *Education Student Asst.*

Andrea Feller, *Education Coordinator*

Amy Frentz, *Curatorial Student Asst.*

Ann Glaros, *Student Education Student Asst.*

Stacie Jackson, *Registrarial Intern*

Stu Mitnik, *Registrarial Volunteer*

Karyn Murphy, *CRC Student Program Coordinator*

Jennifer Passehl, *Registrarial Student Asst.*

Jenea Sanchez, *Business Office Student Asst.*

Janelle Steller, *Administrative Office Intern*

Jettie Vanderveen, *Student Curatorial Asst.*

Christine Zylla, *Student Curatorial Asst.*

ADMINISTRATION Arizona State University

Michael M. Crow, *President*

Elizabeth Capaldi, *Executive Vice President and Provost of the University*

David Young, *Senior Vice President for Academic Affairs*

Gail Hackett, *University Vice Provost and Dean, University College*

Maria T. Allison, *University Vice Provost and Dean, Graduate Studies*

Arthur Blakemore, *Senior Vice Provost*

Mark Searle, *Vice President for Academic Personnel*

Mernoy Harrison, Jr., *Vice President and Provost, Downtown Phoenix campus*

Albert McHenry, *Vice President and Provost, Polytechnic campus*

Marjorie Zatz, *Interim Vice President and Provost, ASU West*

Carol Campbell, *Executive Vice President and Chief Financial Officer*

Christine K. Wilkinson, *Senior Vice President, Secretary of the University, and President of the ASU Alumni Association*

Richard Stanley, *Senior Vice President and University Planner*

Jonathan Fink, *Vice President for Research and Economic Affairs*

Paul Ward, *Vice President for University Administration and Legal Affairs*

James Rund, *Vice President for University Student Initiatives*

Lisa Love, *Vice President for University Athletics*

Anthony "Bud" Rock, *Vice President for Global Engagement*

Adrian Sannier, *Vice President and University Technology Officer*

Virgil Renzulli, *Vice President for Public Affairs*

Eugene E. Garcia, *Vice President for University-School Partnerships*

Herberger College Of The Arts

Kwang-Wu Kim, *Dean*

Dayle Matchett, *Director of Development*

Lynne Aspnes, *Associate Dean and Professor of Music*

Heather Landes, *Assistant Dean for Enrollment and Student Success*

Catherine Richmond, *Director, Fiscal Business Services*

Museum Advisory Board

Eddie Shea, *Chair*

Ridge Smidt, *Executive Committee*

Annie Wacker, *Executive Committee*

Mikki Weithorn, *Executive Committee*

Vanessa Westfall, *Executive Committee*

Beverly Adams

Dorina Bustamante

Erik R. Duncan

Greg Esser

Dr. Renata Hejduk

Scott Jarson

William F. Joffroy, Jr.

Sebastean Loescher

Scott Lyon

Peter Shikany

Marilyn A. Zeitlin, *ex officio*

LaReal Eyring, *ex officio*

Susan Ables, *staff liaison*

Richard Whitney, *honorary member*

Contributors

A graduate of Yale University and the University of Delaware's Winterthur Program in Early American Culture, Ulysses Grant Dietz has been Curator of Decorative Arts at the Newark Museum since 1980. As curator of the museum's 1885 Ballantine House, he was in charge of restoring and reinterpreting this National Historic Landmark in 1994. In 1997 he was the project leader of the first exhibition focused on Newark's once-vast jewelry industry, *The Glitter & The Gold: Fashioning America's Jewelry,* and was the editor of its accompanying catalogue. His latest major publication was *Great Pots: Contemporary Ceramics from Function to Fantasy,* on the museum's extensive collection of studio pottery. Dietz lives in suburban New Jersey with his partner of thirty-one years and their two children.

Peter Held has been the Curator of Ceramics for the Ceramics Research Center since 2002, at Arizona State University Art Museum, Tempe, Arizona. He was the former Director and Curator for the Holter Museum of Art, Helena, Montana. A past resident artist at the Archie Bray Foundation for the Ceramic Arts, he has curated over sixty exhibitions since 1989 including *Ashen Beauty: Woodfired Ceramics* (1990); *David Shaner: A Potter's Work, 1963–1993* (1993); *A Ceramic Continuum: Fifty Years of the Archie Bray Influence* (2001); *Between Clouds of Memory: Akio Takamori, A Mid-Career Survey* (2005) and *Following the Rhythms of Life: The Ceramic Art of David Shaner* (2007). He serves as a board trustee for the American Craft Council and on the advisory boards of Pottery Northwest, Seattle, and the Holter Museum of Art, Helena, Montana. In 2007 he was presented with the CLAY award by the Friends of Contemporary Ceramics for lifetime achievement in the field.

Edward Lebow is an award-winning journalist who has written museum catalogues and articles for *American Craft, Studio Potter,* and *Ceramics: Art and Perception*, among other publications. He contributed a catalogue essay to *Between Clouds of Memory: Akio Takamori,* A Mid-Career Survey, organized in 2005 by the Arizona State University Art Museum Ceramics Research Center. He curated the 1995 retrospective of ceramics by Kenneth Ferguson for the Nelson-Atkins Museum in Kansas City and wrote the catalogue essay for the 1992 survey of Kenneth Price's ceramics at the Menil Collection in Houston, and the Walker Art Center in Minneapolis. More recent contributions appeared in catalogues accompanying *Miró: Playing with Fire* at the George R. Gardiner Museum of Ceramic Arts in Toronto and *Robert Archambeau: Artist, Teacher, Collector* at the Winnipeg Art Gallery. He is the Public Art Program Director for the Phoenix Office of Arts and Culture.

Index

T

U

V

W

Y

Z

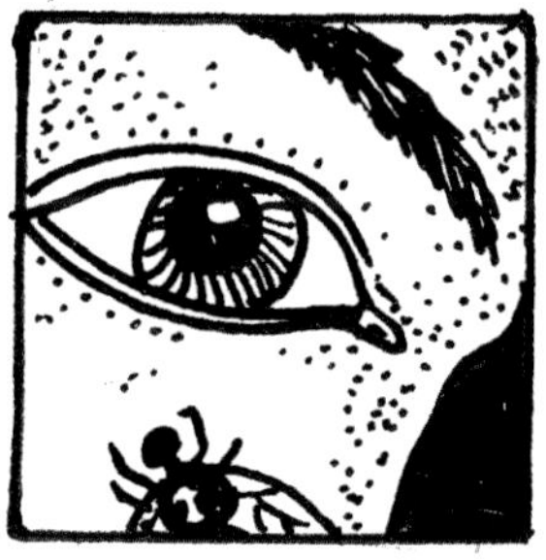